T0328943

Farm animal proteomics 2013

Farm animal proteomics 2013

Proceedings of the 4th Management Committee Meeting and
3rd Meeting of Working Groups 1, 2 & 3 of COST Action FA1002

Košice, Slovakia

25-26 April 2013

edited by:
André de Almeida
David Eckersall
Elena Bencurova
Saskia Dolinska
Patrik Mlynarcik
Miroslava Vincova
Mangesh Bhide

Wageningen Academic
P u b l i s h e r s

ISBN: 978-90-8686-222-1
e-ISBN: 978-90-8686-776-9
DOI: 10.3920/978-90-8686-776-9

Cover drawing by Simão Mateus

First published, 2013

© Wageningen Academic Publishers
The Netherlands, 2013

The Cost Organisation

COST – the acronym for European Cooperation in Science and Technology – is the oldest and widest European intergovernmental network for cooperation in research. Established by the Ministerial Conference in November 1971, COST is presently used by the scientific communities of 35 European countries to cooperate in common research projects supported by national funds.

The funds provided by COST – less than 1% of the total value of the projects – support the COST cooperation networks (COST Actions) through which, with EUR 30 million per year, more than 30,000 European scientists are involved in research having a total value which exceeds EUR 2 billion per year. This is the financial worth of the European added value, which COST achieves.

A 'bottom up approach' (the initiative of launching a COST Action comes from the European scientists themselves), 'à la carte participation' (only countries interested in the Action participate), 'equality of access' (participation is open also to the scientific communities of countries not belonging to the European Union) and 'flexible structure' (easy implementation and light management of the research initiatives) are the main characteristics of COST.

As precursor of advanced multidisciplinary research COST has a very important role for the realisation of the European Research Area (ERA) anticipating and complementing the activities of the Framework Programmes, constituting a 'bridge' towards the scientific communities of emerging countries, increasing the mobility of researchers across Europe and fostering the establishment of 'Networks of Excellence' in many key scientific domains such as: Biomedicine and Molecular Biosciences; Food and Agriculture; Forests, their Products and Services; Materials, Physical and Nanosciences; Chemistry and Molecular Sciences and Technologies; Earth System Science and Environmental Management; Information and Communication Technologies; Transport and Urban Development; Individuals, Societies, Cultures and Health. It covers basic and more applied research and also addresses issues of pre-normative nature or of societal importance.

Web: http://www.cost.eu

Editors

- Dr. André de Almeida, PhD., Instituto de Investigação Científica Tropical, Portugal
- Prof. Dr. David Eckersall, PhD., University of Glasgow, United Kingdom
- Mgr. Elena Bencurova, University of veterinary medicine and pharmacy in Kosice, Slovakia
- Dr. Saskia Dolinska PhD., University of veterinary medicine and pharmacy in Kosice, Slovakia
- Mgr. Patrik Mlynarcik, University of veterinary medicine and pharmacy in Kosice, Slovakia
- Mgr. Miroslava Vincova, University of veterinary medicine and pharmacy in Kosice, Slovakia
- MVDr. Mangesh Bhide PhD., University of veterinary medicine and pharmacy in Kosice, Slovakia

Scientific committee for the book of proceedings

- MVDr. Mangesh Bhide, PhD., University of veterinary medicine and pharmacy in Kosice, Slovakia
- Dr. André de Almeida, PhD., Instituto de Investigação Científica Tropical, Portugal
- Prof. MVDr. Juraj Pistl, PhD., University of veterinary medicine and pharmacy in Kosice, Slovakia
- Dr. Jilian Bryce, PhD., University of Glasgow, United Kingdom
- Dr. Pedro Rodrigues, PhD., University of the Algarve, Portugal
- Dr. Kristin Hollung, PhD., Nofima, Norway
- Prof. Dr. David Eckersall, PhD., University of Glasgow, United Kingdom
- Dr. Fabrizio Ceciliani, PhD., University of Milano, Italy
- MVDr. Tomas Csank, PhD., University of veterinary medicine and pharmacy in Kosice, Slovakia

Table of contents

Part II – Advancing methodology for farm animal proteomics and bioinformatics

Part III – Infectious diseases

Part IV – Animal production

Part V – Food safety quality

Farm animal proteomics: going to the European capital of culture 2013

Mangesh Bhide
Laboratory of Biomedical Microbiology and Immunology, University of Veterinary Medicine and Pharmacy in Kosice, Slovakia; mangeshbhide@me.com

It was almost a year ago, that the management committee of COST Action FA1002 decided to take Farm Animal Proteomics from the westernmost point of Europe, the beautiful coastal harbor of *Vilamoura*, to the vibrant central European city - Kosice, also the European Capital of Culture for 2013.

On behalf of the organizers and FA1002, we have great pleasure in welcoming you to the 3rd annual meeting on Farm Animal Proteomics. We are sure that this meeting will once more immerse you all in the excellent research activities. This year, joined by eminent speakers from most Europe countries, as well as Australia and New Zealand. We hope that this meeting will serve as a platform to present and discuss the newest results and build scientific networks and collaborations in the form of scientific literature and research projects, and very importantly, friendship bonds. Following the lectures of invited speakers, we hope that the meeting will allow time for listening and meeting scientists in the earlier stage of their careers, particularly short-term-scientific mission (STSM) grantees. With fourteen key speeches, seventy five extended-abstracts, thirty eight posters and twenty two scientific lecture presentations from young scientists, the event has numerous scientific themes to share, learn and collaborate.

We are sure that the academic and cultural heritage of Kosice will infuse a sense of the pursuit of excellence. Additionally, the architecture and rich history of the city, a legacy of the numerous civilizations and cultures that passed through Kosice will provide a magical charm and truly fairy-tale memories to take back home. Košice, the metropolis of Eastern Slovakia, laying in the valley of the Hornad River in the basin that shares its name, is the regional administrative centre and Eastern Slovakia's hub of industry, commerce, science and culture. Košice is a city with an eventful and illustrious past, its earliest recorded mention dating from 1230, when it is referred to as 'Villa Cassa'. The coat of arms is the oldest in Europe, a fact attested to by a letter dated 1369. The city's historic sights - from various eras - are concentrated in the historic centre, which is an Urban Heritage Area.

We also hope you will enjoy the social event and engage in productive interactions with bouquet, color and taste of *Tokaj* Wine. So, enjoy the gorgeous wine cellars in *Tokaj*, descend into its mysterious silences, and coolness, covered with mold and have a look into dark galleries dug in bare volcanic rock. Listen carefully to the whisper of a thick wine, slowly poured into the glass as it was tired of waiting for so long until some Scientist dares to taste it ... 'A rush of wine to the head' ... You will never forget.

Part I
Invited plenary communications

Mining deeper into the proteome: pros and cons of pre-fractionation and depletion

Ingrid Miller
Institute of Medical Biochemistry, Department of Biomedical Sciences, University of Veterinary Medicine, Vienna, Austria; ingrid.miller@vetmeduni.ac.at

Previous studies on the complexity of proteomes have estimated protein concentrations spanning over at least 10 orders of magnitude for serum/plasma, and of 6-8 orders of magnitude for cells. At the same time, marker proteins characteristic for particular diseases are supposed to be found in the low to trace abundance range. With conventional proteomic methods 2-4 orders of magnitude may be covered, thus making it necessary to look for strategies for digging deeper into the proteome.

For two-dimensional electrophoresis (2-DE) and without changing the method of sample preparation, a first attempt to gain sensitivity could be the use of larger gels, increased amount of loaded protein, application of more sensitive detection methods, as well as focusing on a narrower pH-range in the first dimension.

Pre-fractionation and depletion both change the composition of the original sample, and are additional steps prior to separation. They aim at reducing the complexity of the sample, either by removing high abundance proteins or by enriching proteins with particular properties. This starts with the buffer selection for sample extraction, which can also be performed in multiple steps with solutions of increasing dissolving power. Subcellular fractionation in a density gradient separates the organelles present in a cell, allowing the study of their subproteomes. Separation based on physico-chemical properties of the proteins may be achieved by chromatography. The most commonly applied principles are ion-exchange and reversed-phase chromatography, as well as affinity-based reactions (for instance enrichment of phosphoproteins). Preparative electrophoresis can also be a versatile tool, especially if it applies conditions different to a subsequent one- or two-dimensional separation (for instance, under native conditions, compared to denaturing and reducing 2-DE).

Similar principles are applied in depletion approaches, usually with the aim of removing 'unwanted' highly abundant proteins. For serum, this is often performed for albumin and IgG, but there are an increasing number of commercial products designed for removal of several plasma proteins. Most of these products are developed for human proteins, and need to be carefully tested for their usefulness in animal samples.

It is important to note that those sample pretreatment steps do not always give clearcut separations: a protein may be present in more than one fraction, or it may appear in another fraction when its properties change (for instance, in the course of a disease). Thus, care has to be taken to check and optimize the respective methods for the samples to be investigated.

Examples will be presented from different fields, mainly for 2-DE. A more detailed overview is given in Miller (2011).

References

Miller, I., 2011. Protein Separation Strategies. In: Eckersall, P.D., Whitfield, P.D. (eds.) Methods in Animal Proteomics. John Wiley & Sons, Chichester, UK, pp. 41-76.

Detection and annotation of common post-translational modifications in mass spectrometry data

Julien Mariethoz[1], Oliver Horlacher[1], Frederic Nikitin[1], Matthew P. Campbell[2], Nicolle H. Packer[2], Markus Muller[1], Frederique Lisacek[1]

[1]*Proteome Informatics Group, SIB Swiss Institute of Bioinformatics, Geneva, Switzerland; frederique.lisacek@isb-sib.ch*
[2]*Biomolecular Frontiers Research Centre (BFRC), Macquarie University, Sydney, Australia*

Objectives

Glycosylation is probably the most important post-translational modification in terms of the number of proteins modified and the diversity generated. In spite of such a central role in biological processes, the study of glycans remains isolated, protein-carbohydrate interactions are rarely reported in bioinformatics databases and glycomics is lagging behind other -omics. Recent progress in method development for characterising the branched structures of complex carbohydrates has now enabled higher throughput technology. Automation then calls for software development. Adding meaning to large data collections requires corresponding bioinformatics methods and tools. Current glycobioinformatics resources do cover information on the structure and function of glycans, their interaction with proteins or their enzymatic synthesis. However, this information is partial, scattered and often difficult to get to for non-glycobiologists.

Phosphorylation is essential for the regulation of cellular processes. The CID-MS/MS fragmentation patterns of phosphopeptides differ from those of the non-phosphorylated peptides and MS/MS search engines often do not consider these differences. Further, even when the true peptide was found, the correct positioning of the phosphate on a residue is hampered by neutral loss of the phosphate, incomplete fragmentation, and coelution of peptides with different phosphate attachment sites.

Material and methods

In partnership with expert international research groups we are involved with the development of the UniCarb KnowledgeBase (UniCarbKB), an effort to develop and provide an informatic framework for the storage and the analysis of high-quality data collections on glycoconjugates. UniCarbKB (http://unicarbkb.org) is an initiative designed to support research in systems biology by complementing proteomics with glycomics (Campbell *et al.*, 2011). It aims to: (1) organise data to enable user-friendly interaction and querying by adopting standardisation and controlled vocabulary guidelines; (2) build a platform that will support the inclusion of new data mining tools and connect disparate existent glycobiology resources; (3) integrate functional data through cross-linking with sugar-binding information. The framework adopts agreed standards to store structural and metadata content including the translation of GlycoSuiteDB structure entries (Cooper *et al.*, 2003) into the GlycoCT format (Herget *et al.*, 2008) offering a comprehensive structure database.

In order to tackle phosphate positioning issues we developed a new algorithm, which takes the neutral loss peaks into account and reports a positioning score and false localization rate for all potential phosphorylation sites. The algorithm is self-learning and can adapt itself to new MS/MS data (Horlacher *et al.*, unpublished data).

Results and discussion

UniCarbKB offers a unique approach to access the most comprehensive biocurated overview of existing glycoinformation associated with proteins in a site-specific manner both from the attachment and the recognition perspective. The initiative is driven as a community-endeavour to promote data sharing in glycobiology and ensure its future development and growth. As mass spectrometry has become the most common method for solving glycan structures and identifying glycopeptides, there is now a substantial range of software tools that are available for analysing MS data produced in glycomics. Tools integrated in the UniCarbKB platform benefit from this work.

Our phosphate-positioning algorithm outperforms the AScore algorithm, which is widely used in the proteomic community. On test datasets, our algorithm yielded a false localisation rate (FLR) between 6.1% and 8.8% depending on preset conditions, whereas the AScore method had a localisation error of 16.6%. Overall, AScore proved to be less sensitive at a given FLR. At a FLR of 1% or higher it identified 30% less correct phosphosites. These results need be confirmed using further large-scale phosphopeptide datasets. This algorithm is to be integrated in the QuickMod (Ahrne *et al.*, 2011) platform that processes spectra for the accurate detection of PTMs in MS/MS data and based on annotated spectral libraries.

References

Ahrne, E., Nikitin, F., Lisacek, F. and Muller, M., 2011. QuickMod: A tool for open modification spectrum library searches. J Proteome Res 10: 2913-2921.

Campbell, M.P., Hayes, C.A., Struwe, W.B., Wilkins, M.R., Aoki-Kinoshita, K.F., Harvey, D.J., Rudd, P.M., Kolarich, D., Lisacek, F., Karlsson, N.G. and Packer, N.H., 2011. UniCarbKB: putting the pieces together for glycomics research. Proteomics 11: 4117-4121.

Cooper, C.A., Joshi, H.J., Harrison, M.J., Wilkins, M.R. and Packer, N.H., 2003. GlycoSuiteDB: a curated relational database of glycoprotein glycan structures and their biological sources. 2003 update. Nucleic Acids Res 31: 511-513.

Herget, S., Ranzinger, R., Maass, K. and Lieth, C.W., 2008. GlycoCT-a unifying sequence format for carbohydrates. Carbohydr Res 343: 2162-2171.

Applied bioinformatics in the structural, post-genomic era

Dimitrios Vlachakis, Georgia Tsiliki, Dimitrios Kondos, Dimitrios-Georgios Kontopoulos, Christos Feidakis and Sophia Kossida
Bioinformatics & Medical Informatics Team, Biomedical Research Foundation, Academy of Athens, Soranou Efessiou 4, 11527 Athens, Greece; skossida@bioacademy.gr

Objectives

The sheer size of genomic data is overwhelming and has recently emerged as the major bottleneck in the field of modern bioinformatics. Nowadays, there are huge genomic databases available that remain unexploited due to the lacking of efficient mining algorithms and processing power (Dobson *et al.*, 2004; Illergård K, 2009). Even so, the quest of joining the dots between sequence and biological function is still primarily based on outdated primary sequence comparisons, which do not cover similarity searches adequately. Provided that secondary, tertiary and quaternary structure is far more conserved in proteins than their corresponding primary amino acid sequence, there is a dear need for new structure-based similarity searching approaches (Kolodny *et al.*, 2005). All evolutionary relationships of proteins, protein structure-function predictions, molecular docking and comparative modeling should all be based on analyses, searches and databases containing structural information (Berbalk *et al.*, 2009).

Material and methods

Herein, we report a new tool that we developed towards this direction; to perform protein similarity searches based on protein secondary structural elements rather than just primary sequence information. The query input sequence can either be of known or unknown structure. In both cases the primary amino acid sequence is converted to amino acid Structural Features Sequence (PSSP) format. The PSSP format is 'per residue' annotation method based on the structural conformation of each amino acid in the query sequence (α-helix, β-sheet, coil, etc). If the query input sequence is of unknown structure, it is subjected to secondary structure prediction algorithms and the SFS format is consequently deduced. Our algorithm is broken down in three parts; in part 1, the query sequence is converted to PSSP format, in part 2 the Query-PSSP is structurally aligned against the reference PSSP databases and finally, in part 3, the structural similarity results are combined with classic primary sequence based results and output to user.

Results and discussion

Our methodology aims to provide the tools essential for performing similarity searches amongst proteins using structural rather than primary sequence information. The query input sequence may either be of known or unknown tertiary or quaternary structure (Figure 1). In either case the primary protein sequence must be transformed to amino acid PSSP format. The PSSP format is a residue-annotation method based on the structural conformation of each amino acid in the query sequence. This type of secondary structure annotation has previously been introduced by the DSSP suite.

All residues that form an a-helix will be substituted with an 'H', a beta-sheet with an 'S', and a coil formation with a 'C' until all query residues have been annotated with a PSSP value. In the case, where the query input sequence lacks 3D structural information, using STRAP it is subjected to secondary structure prediction, where the PSSP format will be subtracted. The exact same PSSP formatting methodology with or without secondary prediction algorithm will be applied to the full NCBI Genomic database. On the other hand, due to the fact that all entries in RCSB database encompass secondary structural information, the transformation to PSSP format may be achieved without performing secondary structure predictions. The proposed algorithm is summarized in Figure 1.

Due to the fact that sequence-based data is by default incomplete (in the case of genomic sequences), which means that they lack structural information, a fast and efficient secondary structure prediction algorithm will be generated and applied. Nonetheless, even upon

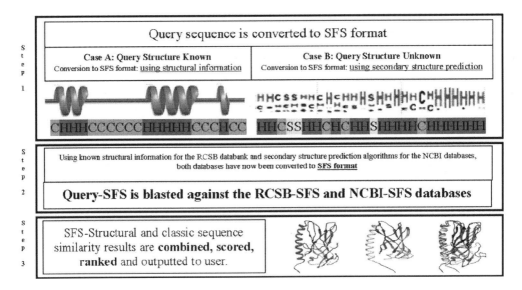

Figure 1. Diagram of the proposed approach.

application of the secondary structure predictions it is still possible to obtain 'unclear' data in the cases where the secondary prediction score do not indisputably indicate a certain secondary structural element. Therefore, the plan is to develop a strategy comprising of two different lines that tackle this issue: First line is the usage of multiple algorithms, which perform secondary structure predictions. The application of a repertoire of different algorithms and approaches on the same sequence string will ensure that eventually the user will end up with a 'consensus prediction', which is statistically more reliable. Second line, is the development of a fast routine, which aims to recognize and in the same time annotate the origin and functionality of each query DNA sequence string using artificial intelligence and machine learning techniques. Then using comparative three dimensional modelling a final prediction of the various structural elements will be made. This way a set of secondary elements will be attributed to the query sequence. Statistically, the weight ratios used by the secondary prediction and homology modelling algorithms will be user defined, optimized for the given needs of the experiment. For instance, obtaining unclear data from a query DNA sequence that has been found to contain conserved features of a designated transcription factor family, which is made up by alpha-helical repeats, will inform and train our algorithm to expect similar patterned alpha-helical conformations too.

Acknowledgements

This research has been co-financed by the European Union (European Social Fund – ESF) and Greek national funds through the Operational Program 'Education and Lifelong Learning' of the National Strategic Reference Framework (NSRF) – Research Funding Program: Thales. Investing in knowledge society through the European Social Fund.

References

Berbalk, C., Schwaiger, C.S. and Lackner, P., 2009. Accuracy analysis of multiple structure alignments. Protein Sci 18: 2027-2035.

Dobson, P.D., Cai, Y.D., Stapley, B.J. and Doig, A.J., 2004. Prediction of protein function in the absence of significant sequence similarity. Curr Med Chem 11: 2135-2142.

Illergård K, A.D., Elofsson A., 2009. Structure is three to ten times more conserved than sequence--a study of structural response in protein cores. Proteins. 77: 499-508.

Kolodny, R., Koehl, P. and Levitt, M., 2005. Comprehensive evaluation of protein structure alignment methods: scoring by geometric measures. J Mol Biol 346: 1173-1188.

PTMomics – a potpourri of experimental approaches

Ana V. Coelho

Instituto de TecnologiaQuímica e Biológica, UniversidadeNova de Lisboa, Oeiras, Portugal;
varela@itqb.unl.pt

The modification of proteins post-translationally is of great importance in protein activity and cellular metabolism regulation. The protein amino acids may be modified after translation, altering the proteins and peptides molecular masses, which can complicate the identification of proteins, in particular from organisms that do not have their genomes sequenced.

There are over 400 post-translational modifications of proteins described. The most common and naturally occurring PTMs include phosphorylation, glycosylation, cleavage, formylation, methionine oxidation and ubiquitination. There are multiple experimental approaches that can be used for their detection and for the characterization and quantification of the generated protein isoforms. In order to achieve these aims it is important to individualize the isoforms either at the protein or at the peptide level. In that respect, 2D electrophoresis has been a helpful separation method, which can be coupled with specific stains or antibody interactions for PTM extensive detection and quantification and with mass spectrometry (MS) for isoform characterization and localization of modified amino acid residues.

Depending on the type of PTM under study different MS strategies tend to be selected mainly due to their lability and chemical specificities. Additionally, in some cases it is indispensable to perform previous enrichment protocols.

Several PTMomics case studies reported by our group and involving PTM identification, characterization and quantification will be discussed (Gomes *et al.*, 2006; Zhenjia *et al.*, 2007; Antunes *et al.*, 2010; Franco *et al.*, 2012).

References

Gomes, R., Vicente Miranda, H., Sousa Silva, M., Graça, G., Coelho, A., Ferreira, A., Cordeiro, C., PoncesFreire, A., 2006. Yeast protein glycation *in vivo* by methylglyoxal: Molecular modification of glycolytic enzymes and heat shock proteins. FEBS J 273:5273-5287

Zhenjia, C., Franco, C.F., Baptista, R.P., Cabral, J.M.S., Coelho, A.V., Rodrigues, C.J., Melo E.P., 2007. Purification and identification of cutinases from *C kahawae* and *C gloesporioides*. Applied Microbiol Biotechnol 73:1306-1313

Antunes, A.M.M., Godinho, A.L.A., Martins, I.L., Oliveira, M.C., Gomes, R.A., Coelho, A.V., Beland, F.A., Marques,M.M., 2010. Protein adducts as prospective biomarkers of nevirapine toxicity. Chemical Res Toxicol 23:1714-1725

Franco, C.F., Soares, R., Pires, E., Santos, R., Coelho,A.V., 2012. Radial nerve cord protein phosphorylation dynamics during starfish arm tip wound healing events. Electrophoresis 33:3764-3778

Gel-free quantitative proteomics approaches, current status

Gabriel Mazzucchelli and Edwin De Pauw
Mass Spectrometry Laboratory – GIGA Proteomics, University of Liege, Liege, Belgium;
Gabriel.mazzucchelli@ulg.ac.be

Two-dimensional polyacrylamide gel electrophoresis (2D PAGE) has been used for the last forty years as a powerful technique to separate proteins and to compare their relative quantities in complex matrices. A major evolution of the technique was brought by the introduction of Differential in Gel Electrophoresis (DIGE) method. The latest evolution allowed up to 3 protein extract co-separations, by using labeling with different fluorescent dyes enabling their selective detection. The quantitation is performed by densitometry measured on gel images and the gel spots of interest are submitted to enzymatic digestion followed by mass spectrometry analysis and database searches for protein identifications.

Meanwhile, the technological evolution of liquid chromatography separations, mass spectrometry techniques, software and bioinformatics allowed gel free differential protein relative abundance analysis of complex protein mixtures to reach suitable power and availability to compete with gel based methods. Multidimensional (ion exchange, reverse-phase or affinity) *nano*-or capillary chromatography in combination with high resolution, high accuracy and fast scanning mass spectrometers, provide exceptional tools that generate unambiguous protein identification and their accurate relative quantification for discovery studies. Specific hardware and software solutions for selective quantification workflow have also been extensively developed.

This lecture will review these recent advances in MS based methodologies for proteomics studies using gel free techniques.

New proteomics strategies applied to clinical studies

Bruno Domon and Sébastien Gallien
Luxembourg Clinical Proteomics Center, Strassen, Luxembourg; bdomon@crp-sante.lu

Advances in proteomics have always been tightly linked to the development of mass spectrometry instrumentation. While for over one decade liquid chromatography/mass spectrometry based proteomics has relied on shotgun approaches, new hypothesis-driven methods have emerged. The latter allows more systematic analyses of peptides (used as surrogate for the proteins) with a high degree of sensitivity. Thus, they facilitate large-scale quantitative clinical studies, aiming at evaluating and validating biomarker panels in bodily fluids, such as blood or urine.

The new hybrid quadrupole-orbitrap mass spectrometer offers unique capabilities in quantitative proteomics, including high-resolution, accurate mass measurements, and fast acquisition. These features, combined with stable isotope dilution, have enabled the emergence of a novel quantification methods to detect and quantify large sets of peptides in complex mixture. Quantification in clinical samples has been performed either on the precursor ions or on the fragment ions.

In order to perform precise quantification, accounting for the variations occurring during the sample preparation and the LC-MS analysis, isotopically labeled internal standards are systematically used. The approach was applied to precisely quantify potential biomarkers of bladder cancer recurrence in urine, and lung cancer markers in blood samples. The analyses were carried out on a quadrupole-orbitrap mass spectrometer operating in full scan (SIM) and MS/MS modes, in conjunction with polypeptides internal standards. To evaluate the performance of the method, the isotopically labeled internal standards were spiked in the clinical samples before and after the proteolysis. Based on calibration curves of the reporter, the qualified markers were quantified in the low amol concentration range. These results demonstrate proof-of-principle and allow assessing the robustness of the method by benchmarking the results with SRM analyses performed on a triple quadrupole instrument. The data stress the benefit of using the high resolution and accurate mass to increase the selectivity of quantitative assays in very complex matrices, through reduction of signal interferences.

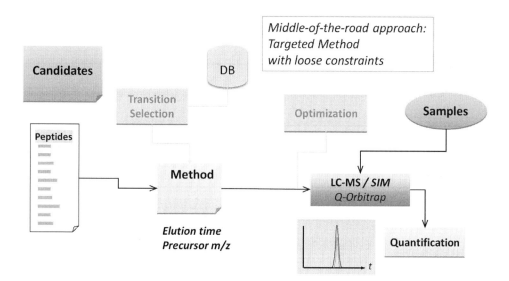

Figure 1. High-resolution / accurate mass workflow applied to quantitative proteomics.

Neuro-immune proteomic crosstalk in health and disease: partners in love, partners in divorce

Norbert Zilka, Zuzana Kazmerova and Michal Novak
Institute of Neuroimmunology, Slovak Academy of Sciences, Bratislava, Slovak Republic
Axon Neuroscience GmbH, Grosslingova 45, Bratislava, Slovak Republic

The permanent neuro-immune dialogue is required for normal brain function. Both systems use the same molecular language; they are covered with similar receptors for neurotransmitters, cytokines, chemokines and trophic factors. Neuronal cells actively participate in the brain immune response whereas glial cells are involved in neurotransmission. In disease condition, this dialogue is significantly impaired, the neuro-immune signaling is not properly regulated, neurons can either overactivate glial cells or on the other hand they can downregulate glial immune response. The abnormal signaling can be further modulated by genetic background, which can explain the different sensitivity of patients to human neurodegenerative disorders. In our study we focused on the 'ON' and 'OFF' signaling which is widely used by neurons to switch on or switch off the glial activity. The lecture will act as a guided tour through the highly variable neuro-immune kingdom, where sometimes partners can live in peace but sometimes they do not understand each other.

Acknowledgements

This work was supported by Axon Neuroscience and structural fund 2624022004 and APVV-0206-11.

Top-down proteomics: 2D gels are an integral part of the process

Jens R. Coorssen

Department of Molecular Physiology, and Molecular Medicine Research Group, School of Medicine, University of Western Sydney, Locked Bag 1797, Penrith South DC, NSW 1797, Australia; j.coorssen@uws.edu.au

Two-dimensional gel electrophoresis (2DE) remains widely regarded as a gold-standard for proteomic analyses. Nonetheless, issues with the method have been routinely noted in the 'review' literature, although there has been little to substantiate these claims and many do not seem plausible when the 2DE technique is considered from the perspective of its underlying chemistry and that of proteins. As (or perhaps *because*) gel-based proteomics is a 'mature' technology, factors contributing to possible reductions in performance are known and thus it is possible to better optimize ongoing analyses by targeted refinement of the technique; in contrast, issues with other approaches that have been popularised over the last decade are only now being more widely recognised. Here I briefly review efforts from my group to quantitatively improve every stage of 2DE analysis, from sample preparation and protein extraction, to in-gel spot 'fractionation' for further improving overall protein resolution, through to improved in-gel protein detection approaches for the enhancement of total proteome coverage. As the overall objective is to provide optimal analyses of (patho)physiological mechanisms, we have applied this refined 2DE protocol in proteomic investigations of human preterm labour, spinal cord injury, and a number of other conditions/samples relevant to both basic and clinical sciences.

Initially we sought to address issues of sample handling and extraction, and the resolution of hydrophobic proteins using 2DE. We suspected that membrane proteomes were regarded in the literature as difficult to resolve by 2DE due at least in part to poor extraction and the substantial amounts of soluble protein in total extracts. We initially addressed existing issues in six ways. First, we tested for ourselves and adopted specific protocol amendments that were quantitatively well supported by the available literature; in particular, this has included sample alkylation and reduction before both the first and second dimensions of 2DE (Herbert *et al.*, 1998, 2001). Second, stacking gels are routinely used in the second dimension, and this SDS-PAGE step is carried out in a cold room (i.e. 4 °C) to fully optimise protein resolution (Coorssen *et al.*, 2002). Third, detailed analyses established that automated frozen disruption (AFD), in which tissue samples are routinely and reproducibly powdered in the deep frozen state, was the best starting point for protein extraction (i.e. providing the highest possible yields) and that this was particularly advantageous for samples that were widely regarded as difficult to handle (e.g. muscle and plant tissues) (Butt and Coorssen, 2006). Fourth, as the least disruptive of prefractionation strategies, we introduced a simple physical separation, using hypotonic lysis and ultracentrifugation to fractionate samples into total soluble and total membrane proteins; each is then resolved separately in our proteomic analyses (Butt and

Coorssen, 2005). Fifth, complementing the minimal prefractionation, postfractionation using third electrophoretic separations enabled the critical resolution of proteins obscured due to stacking at the pH extremes of the 2DE gel, migration at the resolving front, or co-migration with proteins of higher abundance; proteome coverage was thus markedly enhanced using nothing more complicated than an additional SDS-PAGE resolving gel (Butt and Coorssen, 2005). Sixth, as many membrane proteins are influenced by local membrane structure, we also tested a variety a native membrane lipids as adjuncts to the standard CHAPS used in most 2DE extraction buffers. Notably, we found that adding lysophosphatidylcholine (LPC) significantly improved the extraction of membrane proteins, and that this was particularly effective with neuronal tissues (Churchward et al., 2005). Quantitative assessments have also highlighted the importance of establishing the quality of commercial IPG (immobilised pH gradient) strips, particularly with regard to the analysis of membrane proteomes (Taylor and Coorssen, 2006). The end result has been the highest quality, quantitative analyses of soluble and membrane proteomes from a variety of tissues, establishing that high resolution top-down analyses – even of hydrophobic and membrane microdomain proteomes – can be easily and routinely done using 2DE (Butt and Coorssen, 2005; 2006; Butt et al., 2006; Churchward et al., 2005; Furber et al., 2009, 2010; Rogasevskaia and Coorssen, 2011).

In contrast to the dogma found in the review literature, our studies have established that the main issue with 2DE is unlikely to be one of protein resolution, but rather of protein detection (Coorssen, 2012; Gauci et al.). Our ongoing work has thus focussed on detailed analyses of existing stain formulations and the development of better detection tools. Despite some claims, relative to Sypro Ruby (SR) we found no improvement in protein detection using some of the newer stain formulations (i.e. Krypton, Deep Purple, Rubeo, and Flamingo) but did quantitatively establish that Coomassie Brilliant Blue (CBB), when detected using near-IR fluorescence imaging (rather than traditional densitometry) provided protein detection comparable to that of SR (Harris et al., 2007). An extensive comparison of SR and various CBB formulations identified two that were competitive with SR across several criteria, as well as a rapidly saturating SR fluorescence response; thus, while SR detected a small number of additional proteins (i.e. 0.6% of the detected proteome) due to this characteristic, proteome assessment using CBB InfraRed Fluorescence Detection (IRFD) provided the more quantitative analysis of protein changes (and thus lower probability of false-positive and – negative identifications) (R.H. Butt and J.R. Coorssen, in review). With improvements to the formulation and the staining and wash steps, we have now established a CBB-IRFD protocol that is markedly superior to SR, detecting ~20-30% more unique protein spots in both soluble and membrane proteomes resolved by 2DE, but at a fraction of the cost (Gauci et al., 2013). As this new CBB-IRFD protocol is also compatible with subsequent mass spectrometric analyses for protein identification it further enables full top-down proteomic analyses across the entire pI and molecular weight range resolved.

Thus, 2DE is a rigorous, high-resolution technique for large-scale, top-down proteomic analyses, including the dissection of molecular mechanisms. Nevertheless, considering the complexity of native proteomes, it must be remembered that there is no panacea, only pros and cons in all experimental methods. Thus critical, quantitative methodological evaluation and re-evaluation will always lie at the core of the most effective proteomic analyses. It is likely the effective integration of the growing body of available data from all techniques currently in use that will prove the most useful in fully understanding phenotypes (Coorssen, 2012).

References

Butt, R.H. and Coorssen, J.R., 2005. Postfractionation for enhanced proteomic analyses: routine electrophoretic methods increase the resolution of standard 2D-PAGE. J Proteome Res 4: 982-991.

Butt, R.H. and Coorssen, J.R., 2006. Pre-extraction sample handling by automated frozen disruption significantly improves subsequent proteomic analyses. J Proteome Res 5: 437-448.

Butt, R.H., Lee, M.W., Pirshahid, S.A., Backlund, P.S., Wood, S. and Coorssen, J.R., 2006. An initial proteomic analysis of human preterm labor: placental membranes. J Proteome Res 5: 3161-3172.

Churchward, M.A., Butt, R.H., Lang, J.C., Hsu, K.K. and Coorssen, J.R., 2005. Enhanced detergent extraction for analysis of membrane proteomes by two-dimensional gel electrophoresis. Proteome Sci 3: 5.

Coorssen, J.R., 2012. Proteomics. *In: Maloy, S. and Hughes, K. (ed.) Brenner's Encyclopedia of Genetics. Academic Press, Elsevier,*

Coorssen, J.R., Blank, P.S., Albertorio, F., Bezrukov, L., Kolosova, I., Backlund, P.S., Jr. and Zimmerberg, J., 2002. Quantitative femto- to attomole immunodetection of regulated secretory vesicle proteins critical to exocytosis. Anal Biochem 307: 54-62.

Furber, K.L., Churchward, M.A., Rogasevskaia, T.P. and Coorssen, J.R., 2009. Identifying critical components of native Ca2+-triggered membrane fusion. Integrating studies of proteins and lipids. Ann N Y Acad Sci 1152: 121-134.

Furber, K.L., Dean, K.T. and Coorssen, J.R., 2010. Dissecting the mechanism of Ca2+-triggered membrane fusion: probing protein function using thiol reactivity. Clin Exp Pharmacol Physiol 37: 208-217.

Gauci, V.J., Padula, M.P. and Coorssen, J.R., 2013. *Coomassie blue staining for high sensitivity gel-based proteomics.* Journal of Proteomics.

Gauci, V.J., Wright, E.P. and Coorssen, J.R., 2011. Quantitative proteomics: assessing the spectrum of in-gel protein detection methods. J Chem Biol 4: 3-29.

Harris, L.R., Churchward, M.A., Butt, R.H. and Coorssen, J.R., 2007. Assessing detection methods for gel-based proteomic analyses. J Proteome Res 6: 1418-1425.

Herbert, B., Galvani, M., Hamdan, M., Olivieri, E., MacCarthy, J., Pedersen, S. and Righetti, P.G., 2001. Reduction and alkylation of proteins in preparation of two-dimensional map analysis: why, when, and how? Electrophoresis 22: 2046-2057.

Herbert, B.R., Molloy, M.P., Gooley, A.A., Walsh, B.J., Bryson, W.G. and Williams, K.L., 1998. Improved protein solubility in two-dimensional electrophoresis using tributyl phosphine as reducing agent. Electrophoresis 19: 845-851.

Rogasevskaia, T.P. and Coorssen, J.R., 2011. A new approach to the molecular analysis of docking, priming, and regulated membrane fusion. J Chem Biol 4: 117-136.

Taylor, R.C. and Coorssen, J.R., 2006. Proteome resolution by two-dimensional gel electrophoresis varies with the commercial source of IPG strips. J Proteome Res 5: 2919-2927.

MALDI imaging mass spectrometry: applications, limitations and potential

Charles Pineau

Inserm U1085 – IRSET, Proteomics Core Facility Biogenouest, Campus de Beaulieu, University of Rennes I, CS2407, 35042 Rennes cedex, France; charles.pineau@inserm.fr

Matrix assisted laser desorption/ionization (MALDI) tissue imaging mass spectrometry (IMS), introduced in 1997 and exemplified by the work of Caprioli *(Caprioli et al.,* 1997), Stoeckli (Stoeckli *et al.,* 2001) and Chaurand (Chaurand *et al.,* 2002; Chaurand *et al.,* 1999), is a blooming field among the numerous applications of mass spectrometry for protein identification and analysis. IMS is a powerful technique that combines the multichannel (m/z) measurement capability of a mass spectrometer with a surface sampling process that allows probing and analyzing of the spatial arrangement of a wide range of molecules including proteins, peptides – both endogenous or enzymatically produced – lipids, drugs and metabolites, directly from thin slices of tissue (for a review, see Seeley and Caprioli, 2011). Specific information on the relative abundance and spatial distribution of target molecules is maintained, providing the opportunity to correlate ion-specific images with histological features observed by light microscopy. IMS is increasingly recognized as a powerful approach in clinical proteomics, particularly in cancer research (McDonnell *et al.,* 2010). The technology holds a high potential for the discovery of new tissue biomarker candidates, for classification of tumors, early diagnosis or prognosis but also for elucidating pathogenesis pathways and for therapy monitoring (Seeley and Caprioli, 2011).

IMS is also undoubtedly advantageous for assessing the distribution and metabolism of drug candidates within targeted organ sections, including whole-body tissue sections (Cornett *et al.,* 2008; Prideaux *et al.,* 2010). IMS thus offers great potential for biodistribution studies where the technology has two major advantages over whole-body autoradioluminography as it does not require radiolabelling of the drug of interest and can distinguish between the drug and its metabolites (Khatib-Shahidi *et al.,* 2006; Prideaux *et al.,* 2010). Recently, there has also been considerable interest in adding absolute quantitative capabilities to IMS experiments (Signor *et al.,* 2007). Thus, the technology promises to have a major impact in the field of mechanistic and reglementary toxicology in the coming years. Indeed, ongoing technological developments should rapidly lead to powerful and straightforward methods for accurately monitoring toxicants action on health and help understand the mechanistic action of toxicants.. Finally, three dimensional-volume reconstruction MALDI IMS represents the natural extension to the planar IMS (Crecelius *et al.,* 2005). This possibility currently attracts great attention as the ability to spatially and quantitatively correlate biomolecules information with *in vivo* anatomical imaging provided for example by magnetic resonance has tremendous implications (Andersson *et al.,* 2008; Sinha *et al.,* 2008).

References

Andersson, M., Groseclose, M.R., Deutch, A.Y. and Caprioli, R.M., 2008. Imaging mass spectrometry of proteins and peptides: 3D volume reconstruction. Nat Methods 5: 101-108.

Caprioli, R.M., Farmer, T.B. and Gile, J., 1997. Molecular imaging of biological samples: localization of peptides and proteins using MALDI-TOF MS. Anal Chem 69: 4751-4760.

Chaurand, P., Schwartz, S.A. and Caprioli, R.M., 2002. Imaging mass spectrometry: a new tool to investigate the spatial organization of peptides and proteins in mammalian tissue sections. Curr Opin Chem Biol 6: 676-681.

Chaurand, P., Stoeckli, M. and Caprioli, R.M., 1999. Direct profiling of proteins in biological tissue sections by MALDI mass spectrometry. Anal Chem 71: 5263-5270.

Cornett, D.S., Frappier, S.L. and Caprioli, R.M., 2008. MALDI-FTICR imaging mass spectrometry of drugs and metabolites in tissue. Anal Chem 80: 5648-5653.

Crecelius, A.C., Cornett, D.S., Caprioli, R.M., Williams, B., Dawant, B.M. and Bodenheimer, B., 2005. Three-dimensional visualization of protein expression in mouse brain structures using imaging mass spectrometry. J Am Soc Mass Spectrom 16: 1093-1099.

Khatib-Shahidi, S., Andersson, M., Herman, J.L., Gillespie, T.A. and Caprioli, R.M., 2006. Direct molecular analysis of whole-body animal tissue sections by imaging MALDI mass spectrometry. Anal Chem 78: 6448-6456.

McDonnell, L.A., Corthals, G.L., Willems, S.M., van Remoortere, A., van Zeijl, R.J. and Deelder, A.M., 2010. Peptide and protein imaging mass spectrometry in cancer research. J Proteomics 73: 1921-1944.

Prideaux, B., Staab, D. and Stoeckli, M., 2010. Applications of MALDI-MSI to pharmaceutical research. Methods Mol Biol 656: 405-413.

Seeley, E.H. and Caprioli, R.M., 2011. MALDI imaging mass spectrometry of human tissue: method challenges and clinical perspectives. Trends Biotechnol 29: 136-143.

Signor, L., Varesio, E., Staack, R.F., Starke, V., Richter, W.F. and Hopfgartner, G., 2007. Analysis of erlotinib and its metabolites in rat tissue sections by MALDI quadrupole time-of-flight mass spectrometry. J Mass Spectrom 42: 900-909.

Sinha, T.K., Khatib-Shahidi, S., Yankeelov, T.E., Mapara, K., Ehtesham, M., Cornett, D.S., Dawant, B.M., Caprioli, R.M. and Gore, J.C., 2008. Integrating spatially resolved three-dimensional MALDI IMS with *in vivo* magnetic resonance imaging. Nat Methods 5: 57-59.

Stoeckli, M., Chaurand, P., Hallahan, D.E. and Caprioli, R.M., 2001. Imaging mass spectrometry: a new technology for the analysis of protein expression in mammalian tissues. Nat Med 7: 493-496.

Parasite- and host-derived proteins involved in African trypanosome brain invasion and dysfunction

Krister Kristensson
Department of Neuroscience, Karolinska Institutet, Stockholm, Sweden

Human African trypanosomiasis (sleeping sickness) is caused by subspecies of the extracellular parasite *Trypanosoma brucei,* which are spread by the tsetse flies in sub-Saharan Africa. A major pathogenic event, causing a number of neurological dysfunctions including pain and dysregulation of sleep, is the invasion of the parasite into the brain. This neuroinvasion make the disease almost always fatal if untreated. It occurs in two phases: first through fenestrated vessels in the choroid plexus, circumventricular organs and peripheral nerve root ganglia, and then as a multistep process across post-capillary venules in the brain parenchyma. We here present our current understanding on the complex cellular and molecular host-parasite interactions, which underlie the passage of the trypanosomes across the blood-brain barrier (BBB). *In vitro* studies have suggested that phosphatases and proteases expressed on *T. brucei* external surfaces facilitate their passage across the BBB, either via the paracellular or the transcellular route. In a series of experiments we have made the unexpected observation that immune response molecules released as a defense against the parasite play a paradoxical role also to increase neuroinvasion both of the trypanosomes and T cells. This involves molecules release both by the innate and the adaptive immune responses in their interactions. In spite of causing a long-standing inflammation in the brain, there are only minor signs of neurodegeneration. We will discuss experiments aiming to understand how molecules released in the nervous system on the one hand may control trypanosome growth but at the same time be neuroprotective. The neuronal dysfunctions, manifested as disturbances in the sleep pattern may be caused either by trypanosome-derived molecules, e.g. prostaglandins, and/ or immune response molecule released into areas of the nervous system involved in sleep-wakefulness regulation and pain. The need for better understanding of the neurobiology of the pathogenesis of these interactions in order to improve the diagnosis and management of sleeping sickness will be highlighted.

Proteomics at the host: pathogen interface

Richard Burchmore
Institute of Infection, Immunity and Inflammation, College of Medical, Veterinary & Life Sciences, University of Glasgow, United Kingdom; richard.burchmore@glasgow.ac.uk

The surface membrane of cells is the interface with the environment. The lipid membrane is impermeable, so the uptake of molecules by the cell must be facilitated by the action of integral membrane proteins. These proteins, which include transporters, channels and receptors, mediate communication between the cell and the environment and are thus key to all cellular processes. In pathogens, surface membrane proteins represent the interface with the host and the primary route by which drugs can access their targets. Understanding the expression, structure and function of pathogen membrane proteins is central to understanding host:pathogen interactions and offers many possibilities for control of infection.

However, integral membrane proteins are under-represented in proteomic analyses, because they are of relatively low abundance and have biophysical properties that make them difficult to solubilise, digest and ionize. We are interested in the development of approaches to enhance proteomic coverage of integral membrane proteins, including enrichment by subcellular fractionation, selective solubility and surface labelling, and in the establishment of optimised conditions for proteomic analysis of hydrophobic proteins. We have combined these approaches with functional assays for membrane protein activities to understand the contribution of membrane proteins to phenotypes such as virulence and drug resistance in a variety of pathogens.

Microbial proteomics in food safety and animal welfare

Paola Roncada[1,2], Alessio Soggiu[2], Cristian Piras[2] and Luigi Bonizzi[2]
[1]Istituto Sperimentale Italiano L. Spallanzani, Milano, Italy; paola.roncada@guest.unimi.it
[2]DIVET, Dipartimento di Scienze Veterinarie e Sanità Pubblica, Università degli Studi di Milano, Italy

It is mandatory for prevention and control of infectious diseases to have facilities that are able to quickly produce reliable, highly specific and sensible tools that allow on one hand an adequate sanitary surveillance (diagnosis) and on the other to obtain effective operative tools (vaccines, therapeutic molecules). From this point of view, proteomics constitutes a very important approach to integrate with the prevention and control of infectious diseases and in particular of the sanitary emergencies and food safety linked to animal health. In this context, microbial proteomics becomes the hard-core junction made by the thematic nodes of sanitary emergencies for human health (Piras *et al.*, 2012). Prevention of human and animal pathologies depends on the assessment of epidemiologic risks, on earliness of diagnosis and on availability of exposition markers. Such aspects are of particular importance for those pathologies with zoonotic characteristics that are vehiculated by food. In Europe, 60-137 cases/100,000 inhabitants are estimated and the EU has given indications for the improvement of animal derived food sanity. Microbial Proteomics is opening up new possibilities in the study of disease pathogenesis, in animal welfare, in novel diagnostic and therapeutic markers and in risk assessment (Figure 1) (Turk *et al.*, 2012). This powerful tool allows the application of results into the productive chain, with tangible falls on the quality of production and on consumer health care. Microbial protemics in the field of diagnostic and of development of therapeutic molecules and vaccines allows detection of protein targets for early diagnosis and, through the study of protein interactions and of specific receptors, it also allows the identification of targets for the development of vaccines and therapeutic molecules. Microbial proteomics is one of the best tools to control emerging diseases and zoonoses to improve human health and welfare. Different techniques, such as MALDI biotyper, immunoproteomics, metabonomics, integrated together are fundamental for obtaining clear results useful for public health.

References

Piras, C., Soggiu, A., Bonizzi, L., Gaviraghi, A., Deriu, F., De Martino, L., Iovane, G., Amoresano, A. and Roncada, P., 2012. Comparative proteomics to evaluate multi drug resistance in Escherichia coli. Molecular BioSystems 8: 1060-1067.

Turk, R., Piras, C., Kovačić, M., Samardžija, M., Ahmed, H., De Canio, M., Urbani, A., Meštrić, Z.F., Soggiu, A. and Bonizzi, L., 2012. Proteomics of inflammatory and oxidative stress response in cows with subclinical and clinical mastitis. Journal of proteomics.

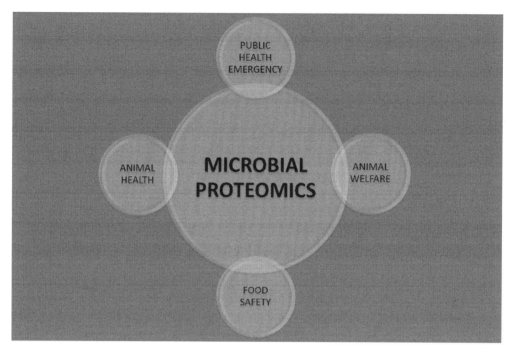

Figure 1. Microbial proteomics network.

The search of stress markers in porcine by using proteomics

Anna Marco-Ramell, Laura Arroyo and Anna Bassols
Departament de Bioquímica i Biologia Molecular. Facultat de Veterinària. Universitat Autònoma de Barcelona. Spain; anna.bassols@uab.cat

Welfare problems are important for ethical reasons and because they may cause great economic losses. Nevertheless, objective laboratorial criteria to evaluate animal stress are still lacking. Several plasma components have been proposed as stress biomarkers, but there are inherent problems to their practical use, due to difficult interpretation of results, lack of specificity, or insufficient number of validation studies. For example, cortisol is accepted as the main marker for stress, but its release is episodic, giving way to a high inter- and intraindividual variation (Mormede *et al.*, 2007). Thus, new approaches are needed to address this question.

We have applied proteomic techniques to the animal stress/welfare problem, with the aim to identify new biomarkers with potential application in animal production and veterinary medicine. Here our last results on pigs will be presented.

In a first study, growing pigs were subjected to changes in housing density with the goal of identifying the best biomarkers for the stress associated to this condition (Marco-Ramell *et al.*, 2011). Eight Duroc × (Landrace × Large White) male pigs housed at a stocking rate of 0.50 m2/pig were subjected to a higher stocking rate of 0.25 m^2/pig (higher density, HD) for two 4-day periods over 26 days. The 15-day interval before the first space restriction was considered an appropriate period to reduce the psychological and physiological effects associated with transport and new housing. A biochemical approach showed that HD housed pigs showed significant differences in total cholesterol and low density lipoprotein-associated cholesterol, as well as in concentrations of the pig-major acute phase protein (Pig-MAP). HD-individuals also showed an imbalance in redox homeostasis, detected as an increase in the level of oxidized proteins measured as the total plasma carbonyl protein content with a compensatory increase in the activity of the antioxidant enzyme glutathione peroxidase. The DIGE methodology was used to identify serum protein biomarkers taking as working samples those corresponding to low-density and high-density conditions. A new putative biomarker for stress, the component of the cell cytoskeleton actin, was identified and subsequently validated by western blot. Its finding in serum samples from high-density housing suggests that a certain degree of cellular damage exists in this situation. The presence of actin in serum samples was validated in all animals by Western blotting, which allowed us to identify the protein as a spot with an expected molecular mass of 42 kDa. Actin has been frequently described as differentially expressed in proteomic studies, and proposed as a general marker for cell damage. In the present study, there was no correlation between actin and the muscle-specific enzyme creatine kinase (CK), indicating that the presence of actin was not specifically related to skeletal muscle damage.

Using a statistical approach widely used in genomics, i.e. data clustering using a software tool developed for this purpose, animals segregated into two main groups. A subset of animals responded with an acute phase-like reaction at several degrees with increased APPs and oxidative stress markers and decreased cortisol, with similarities to a pro-inflammatory situation. This may be similar to the aseptic proinflammatory state proposed in humans to link a stressful way of life, oxidative stress and cardiovascular disease. In other animals, the response appeared to be mainly mediated by cortisol, with the subsequent lipid mobilization, and some signals of cell damage. Our results are in agreement with previous research that has shown that animals adopt different behavioural strategies in order to cope with stress. So-called 'active copers' tend to adopt a fight–flight type of response when challenged, with low HPA-axis reactivity and high activation of the sympathetic-adrenomedullary system. In contrast, 'passive coping' involves immobility, higher activation of the HPA-axis and higher parasympathetic reactivity.

The second stress condition referred to gilts, which are usually housed in groups until they are moved into the insemination room, where they are housed in small and individual stalls. The change to this individual housing system could be stressful for the animals. The European directions have recently been modified to avoid this condition, but we have used it as a model of animal stress. We measured health, nutritional and oxidative stress markers, cortisol and acute phase proteins, with enzymatic or colorimetric assays, on an Olympus automatic analyzer, or with commercial ELISAs. The superoxide dismutase (SOD) and glutathione peroxidase (GPx) activities were measured in erythrocyte lysate, and total glutathione (GSH) in whole blood. SOD, GPx and GSH increased the following day after the housing change in both groups, but they slightly decrease in the study group, but not in the control group. The proteomic approach was performed with 2-DE DIGE and MS and allowed the identification of several proteins, mainly belonging to the acute phase protein category. The results of MS identification revealed that two positive acute phase proteins, haptoglobin and the inter-alpha-trypsin inhibitor heavy chain H4 (ITIH4 or Pig-MAP), modified their levels when the animals were housed in individual boxes. We validated haptoglobin and Pig-MAP with a colorimetric assay and a commercial ELISA, respectively. Both increased on day 3, but they decreased on the following days in the two animal groups. We also measured another acute phase protein, the protein C-reactive (CRP), which increased markedly in the study group, but not in the control one. A quantitative approach was attempted by using ITRAQ, yielding other potential biomarkers. These samples were also useful to validate an SRM strategy to quantify several acute phase proteins that were identified before.

Finally, we have attempted also to understand how the brain is involved in the control of stress and emotion in pigs. The organization of the response to a stressful situation involves the activity of different types of neurotransmitter systems in several areas of the limbic system. Thereby, changes in neurotransmitter (NT) concentrations are related to the activation and modulation of behavioral processes and autonomic response (Mora *et al.*, 2012). For this purpose, several cerebral areas were dissected at the slaughterhouse (hippocampus, hypothalamus, amygdala) and immediately frozen in liquid nitrogen. Neurotransmitters (NT) from the dopaminergic (noradrenaline (NA), dopamine (DA), 3,4-dihydroxyphenylacetic acid (DOPAC) and homovanillic acid (HVA)) and serotononergic (5-hydroxyindole-3-acetic acid (5-HIAA) and serotonin (5-HT)) pathways were identified and quantified by HPLC. The different brain areas showed different patterns. Interestingly, changes in the NT profiles were found in pigs that were subjected to a stressful management at their arrival at the slaughterhouse.

In conclusion, a proteomic approach can lead to the identification of potential stress and welfare markers useful for animal production and management and, in conjunction with other analytical tests, it provides a global vision of the physiological status and the adaptation of the animal to challenging living conditions.

Acknowledgements

Thanks are due to X. Manteca, JL Ruiz de la Torre, A. Velarde, E. Mainau and R Peña for their valuable help in the experimental design and animal management and welfare assessment. This work was supported by Grants AGL2006-02365, AGL2010-21578-C03-03 and AGL2011-30598-C03-02 from the Spanish Ministerio de Ciencia y Tecnologia, and Grant 2009 SGR-1091 from the Generalitat de Catalunya (to A.B.). Part of the funding was financed by the FEDER program from the European Union.

References

Marco-Ramell, A., Pato, R., Peña R., Saco, Y., Manteca, X., Ruiz de la Torre, J.L., Bassols, A. 2011. Identification of serum stress biomarkers in pigs housed at different stocking densities. Veterinary Journal 190: e66-71.

Mora, F., Segovia, G., Del Arco, A., de Blas, M., Garrido, P. 2012. Stress, neurotransmitters, corticosterone and body-brain integration. Brain Research 1476: 71-85.

Mormede, P., Andanson, S., Auperin, B., Beerda, B., Guemene, D., Malmkvist, J., Manteca, X., Manteuffel, G., Prunet, P., van Reenen, C.G., Richard, S., Veissier, I. 2007. Exploration of the hypothalamic-pituitary-adrenal function as a tool to evaluate animal welfare. Physiology and Behavior 92: 317-339.

Quest for biomarkers of the lean-to-fat ratio by proteomics in beef production

Muriel Bonnet[1,2], Nicolas Kaspric[1,2], Brigitte Picard[1,2]
[1]*INRA, UMR1213 Herbivores, 63122 Saint-Genès-Champanelle, France;*
muriel.bonnet@clermont.inra.fr
[2]*VetAgro Sup, Élevage et production des ruminants, 63370 Lempdes, France*

Producing meat animals with adequate muscular and adipose masses (i.e. lean-to-fat ratio) is an economic challenge for the beef industry. In cattle, conformation (muscle mass) and fatness (adipose tissue mass) are evaluated using the European Union beef carcass classification system (EUROP) scale. The EUROP scale determines the price/kg of carcasses. The lean-to-fat ratio is the result of a dynamic balance between the number and size of muscular and adipose cells, respectively (Bonnet *et al.*, 2010). Identifying proteins that contribute to the increase in the number and volume of adipose and muscular cells has implication for the proposition of biomarkers of growth potential and/or of carcass composition.

The total number of muscle fibres is set by the end of the second trimester of gestation (Picard *et al.*, 2002). Conversely, the number of adipocytes is set by birth or by early adulthood, depending on the anatomical location of the adipose tissue (Vernon, 1986). Thus, we hypothesized that a high throughput molecular characterisation of adipose tissue (AT) and muscle from bovine foetuses differing by the age would be a powerful way to identify proteins associated to the increase in the number and the size of muscular and adipose cells.

We combined measurements of chemical composition, cellularity, histology, enzyme activities, gene expression and proteomics to describe the ontogeny of perirenal AT and *Semitendinosus* muscle in bovine at 60, 110, 180, 210 and 260 days post conception (dpc) in Blond d'Aquitaine (n=3 per age) and Charolais (n=5 per age) breeds. These breeds were chosen for their differences in lean-to-fat ratio in the post-natal life.

Between 110 and 260 dpc (38 and 90% of gestation length, respectively), the increase in the weight of perirenal AT resulted from an increase in the volume and mainly in the number of adipocytes (Taga *et al.*, 2011). The increases in adipocyte volume and number were accompanied by changes in the abundance of 128 proteins among the 143 proteins identified and common to the four last fetal ages studied (Taga *et al.*, 2012). Among the identified proteins, some of them have never been described in the AT and may contribute to hyperplasia of adipose precursors, by controlling cell cycle progression, apoptosis and/or by delaying adipocyte differentiation. The age of 180 dpc seems to be a pivotal age for the transition between proliferation and differentiation of adipocyte progenitors. An increase in the abundance of many proteins involved in differentiation and in the increase in adipocyte volume was observed from 180 dpc.

The fetal growth of semitendinosus muscle was accompanied by changes in the abundance of 245 proteins. We revealed high abundance of proteins involved in apoptosis at 60 and 110 dpc, suggesting that the proliferation – apoptosis balance may play a role in the determination of the total number of fibers (Chaze *et al.*, 2008). The age of 180 dpc corresponds to a reduction in cell proliferation and a transition between the formation of myofibres and their maturation. Indeed, the abundance of stathmin, which has an important role in cell cycle regulation, decreased from 180 dpc onwards. Septin proteins (septin 2 and 11 isoforms) involved in cytoskeletal organisation, scaffolding and cell division plane had a stable expression up to 180 dpc, consistent with intense cell division until this stage, and declined thereafter. Annexin A1, which has an anti proliferative function via the activation of ERK pathways, showed increased abundance from 180 dpc onwards. Increased maturation of fibers from 180 to 260 dpc is reflected by significant changes in the profiles of protein isoforms belonging to metabolic and contractile pathways (Chaze *et al.*, 2009).

The cellular and molecular features of AT and muscle during ontogenesis provide potential hallmarks of adipose and muscular cells hyperplasia or hypertrophy. Among these proteins we hypothesize there are 'master' proteins, thus a current integrative bioinformatic analysis of adipose and muscular data aims to identify them. Then, the relationship between the abundance of these proteins and data from carcass composition will be studied in tissues from meat producing cattle differing by their rearing conditions or by their genotypes.

References

Bonnet, M., Cassar-Malek, I., Chilliard, Y. and Picard, B., 2010. Ontogenesis of muscle and adipose tissues and their interactions in ruminants and other species. Animal 4: 1093-1109.

Chaze, T., Meunier, B., Chambon, C., Jurie, C. and Picard, B., 2008. *In vivo* proteome dynamics during early bovine myogenesis. Proteomics 8: 4236-4248.

Chaze, T., Meunier, B., Chambon, C., Jurie, C. and Picard, B., 2009. Proteome dynamics during contractile and metabolic differentiation of bovine foetal muscle. Animal 3: 980-1000.

Picard, B., Lefaucheur, L., Berri, C. and Duclos, M.J., 2002. Muscle fibre ontogenesis in farm animal species. Reproduction, Nutrition, Development 42: 415-431.

Taga, H., Bonnet, M., Picard, B., Zingaretti, M.C., Cassar-Malek, I., Cinti, S. and Chilliard, Y., 2011. Adipocyte metabolism and cellularity are related to differences in adipose tissue maturity between Holstein and Charolais or Blond d'Aquitaine fetuses. Journal of Animal Science 89: 711-721.

Taga, H., Chilliard, Y., Meunier, B., Chambon, C., Picard, B., Zingaretti, M.C., Cinti, S. and Bonnet, M., 2012. Cellular and molecular large-scale features of fetal adipose tissue: is bovine perirenal adipose tissue brown? Journal of Cellular Physiology 227 1688-1700

Vernon, R.G., 1986. The growth and metabolism of Adipocytes. In: P.J. Buttery, D.B. Lindsay and N.B. Haynes (Eds.), Control and Manipulation of Animal Growth. Butterworths, London., pp. 67-83.

The LEXSY platform for recombinant protein expression

Reinhard Breitling

Jena Bioscience GmbH, Jena, Germany; reinhard.breitling@jenabioscience.com

Due to the increasing demand for production of recombinant proteins in the proteomics era and the shortcomings of the traditional protein expression systems we have developed the LEXSY expression platform based on the protozoan organism *Leishmania tarentolae*. This unicellular eukaryotic host is safe (biosafety group S1), robust, easy and cost-efficient to culture and has the full capacity for eukaryotic protein folding and modification including mammalian-type post-translational glycosylation. We demonstrated an exceptionally homogeneous biantennary structure of fully galactosylated, core-α-1,6-fucosylated N-glycans of LEXSY expressed glycoproteins (Figure 1).

Based on the flexibility of the LEXSY technology, recombinant target proteins can be produced intracellularly or be secreted into the culture medium. Both, constitutive and inducible expression architectures are available.

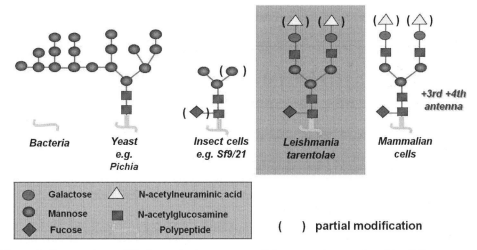

Figure 1. Comparison of N-glycosylation pattern of glycoproteins produced in different protein expression systems. Glycosylation in LEXSY was investigated with human erythropoietin, human interferon gamma and host surface glycoprotein GP63. In all cases a biantennary, fully galactosylated, core-α-1,6-fucosylated N-glycan structure was found that is similar to mammalian-type glycosylation (Breitling et al., 2002).

For constitutive expression, the target gene constructs are stably integrated into the chromosomal rDNA (*ssu*) locus and co-transcribed by the strong RNA polymerase I of the protozoan host cells (Breitling *et al.*, 2002). For inducible expression, we established a *L. tarentolae* recipient strain co-expressing bacteriophage T7 RNA polymerase and tetracycline repressor. This strain was transfected with heterologous target genes placed under the control of a T7 promoter/TET-operator assembly, which allowed transcription initiation upon addition of tetracycline to the culture medium. The target gene constructs were either stably integrated into the chromosomal ß-tubulin (*tub*) or ornithine decarboxylase (*odc*) locus (Kushnir *et al.*, 2005) or maintained episomaly (Kushnir *et al.*, 2011). In a recent version induction of target protein expression can be monitored online during cultivation by a transcriptionally coupled fluorescence marker (www.jenabioscience.com).

Numerous proteins, including enzymes, surface antigens, toxins, antibodies and membrane proteins have been expressed with LEXSY (Table 1). Expression yields of up to several hundred mg per litre of culture were obtained, and the purified proteins were successfully employed for diagnostics and research and development including structure determination by NMR and X-ray crystallography (Niculae *et al.* 2006; Gazdag *et al.* 2010). LEXSY has recently been complemented by a version for *in vitro* translation for cell-free expression of recombinant proteins (Mureev *et al.*, 2009, Kovtun *et al.*, 2010).

Pilot plant cultivations in 30 litre bioreactor scale in standard bacteriological media revealed that LEXSY is fully adapted to fermentation technology. Target protein yields of >100 mg/l were obtained in high cell density fermentations (8×10^8 cells/ml) in suspension culture with this robust protozoan host.

In summary: LEXSY thus combines the advantages of a fast growing robust expression host with the full potential of an eukaryotic protein synthesis/folding/modification machinery and will contribute to meet the challenges of ongoing proteomics initiatives.

Table 1. Typical examples of LEXSY-expressed proteins clustered by type of protein.[1]

Target protein	Size (kDa)	Yield (mg/l)
Cytoplasmic proteins		
SOD1	16	30
EGFP	28	300
SPEE	35	30
p85 of PI3 kinase	85	3
smmyHC	154	1
Nuclear proteins		
T7 RNA Pol	100	1
Secreted proteins		
MHC II-b	30	500
CRP	23	44
SAG1&2	15/31	10
Fc fusion	39	10
MDP1	45	6
Laminin 332	420 (150+135+135)	0.5
Membrane proteins		
EGFP-Rab7 (mb-associated)	52	12
BkrB2-GST (Type III TM7)	55	0.5
PDM9 (Type I)	43	0.2

[1] SOD1 = human Cu/Zn superoxide dismutase; EGFP = enhanced green fluorescent protein of A. victoria; SPEE = human spermidine synthetase; p85 = bovine Phosphoinositide 3-Kinase regulatory subunit α; smmyHC = heavy chain of human smooth muscle myosine; T7 RNA Pol. = RNA polymerase of phage T7 supplied with nuclear localization signal; MHC II-β = human Major Histocom-patibility Complex II β subunit; CRP = human C-reactive protein of pentaxin family; SAG1/2 = surface antigens of Toxoplasma gondii Fc fusion = N-terminal fusion of DNA binding domain to human Fc fragment; MDP1 = human renal dipeptidase 1; Laminin 332 = large heterotrimeric human laminin glycoprotein α3β3γ2; EGFP-Rab7 = EGFP fusion of Ras-associated small GTP-binding protein Rab7 (membrane associated by prenylation); BrkB2-GST = GST fusion of human bradykinin receptor B2 (7TM transmembrane protein); PDM9 = human transmembrane protein with EGF-like and two follistatin-like domains 2 (type I membrane protein N out). For detailed description see www.jenabioscience.com.

References

Breitling, R., Klingner, S., Callewaert, N., Pietrucha, R., Geyer, A., Ehrlich, G., Hartung, R., Müller, A., Contreras, R., Beverley, S.M. and Alexandrov, K. 2002. Non-pathogenic trypanosomatid protozoa as a platform for protein research and production. Protein Expression and Purification 25, 209-218.

Gazdag, E.M., Cirstea, I., Breitling, R., Lukes, J., Blankenfeldt, W. and Alexandrov, K., 2010. Purification and crystallization of human Cu/Zn superoxide dismutase recombinantly produced in the protozoan *Leishmania tarentolae*. Acta Crystallographica F66, 871-877.

Kovtun, O., Mureev, S., Johnston, W. and Alexandrov, K., 2010. Towards the Construction of Expressed Proteomes Using a *Leishmania tarentolae* Based Cell-Free Expression System. PLOS one 5, e14388, doi: http://dx.doi.org/10.1371/journal.pone.0014388.

Kushnir, S., Cirstea, I., Basiliya, L., Lupilova, N., Breitling, R. and Alexandrov, K., 2011. Artificial linear episome-based protein expression system for protozoon *Leishmania tarentolae*. Molecular & Biochemical Parasitology 176, 69-79.

Kushnir, S., Gase, K., Breitling, R. and Alexandrov, K., 2005. Development of an inducible protein expression system based on the protozoan host *Leishmania tarentolae*. Protein Expression and Purification 42, 37-46.

Mureev, S., Kovtun, O., Nguyen, U.T.T. and Alexandrov, K., 2009. Species-independent translational leaders facilitate cell-free expression. Nature Biotechnology 27, 747-752.

Niculae, A., Bayer, P., Cirstea, I., Bergbrede, T., Pietrucha R., Gruen, M., Breitling, R. and Alexandrov, K., 2006. Isotopic labeling of recombinant proteins expressed in the protozoan host *Leishmania tarentolae*. Protein Expression and Purification 48, 167-172.

Part II
Advancing methodology for farm animal proteomics and bioinformatics

Data-independent acquisition strategies for quantitative proteomics

Ute Distler, Jörg Kuharev, Hansjörg Schild and Stefan Tenzer
UMC, Johannes Gutenberg University Mainz, Germany; tenzer@uni-mainz.de

Objectives

In shotgun proteomics, data-dependent precursor acquisition (DDA) is widely used to profile protein components in complex samples. Although very popular, there are some inherent limitations to the DDA approach, such as irreproducible precursor ion selection, under-sampling and long instrument cycle times. Unbiased 'data-independent acquisition' (DIA) strategies try to overcome those limitations. In MS^E, which is supported by Waters Q-TOF instrument platforms, such as the Synapt G2-S, a wide band pass filter is used for precursor selection. During acquisition, alternating MS scans are collected at low and high collision energy (CE), providing precursor and fragment ion information, respectively. Introduction of ion mobility separation (IMS), which provides an additional dimension of separation, leads to an increase of identified peptides and proteins in MS^E workflows. For label-free quantification of ion mobility based MS^E data, we developed a bioinformatics pipeline, ISOQuant, allowing retention time alignment, clustering, normalization, isoform/homology filtering, absolute quantification and report generation. Thus, we are able to reproducibly quantify up to 2,500 proteins in a single LC-MS run. The workflow can be adapted to different kinds of proteomic samples providing a robust platform for DIA label-free proteomics.

Material and methods

Aliquots corresponding to 20 µg protein derived from different cell lysates were digested using a modified FASP protocol (Wisniewski *et al.*, 2009). Briefly, redissolved protein was loaded on a centrifugal filter (VIVACON 500, Sartorius Stedim Biotech, Goettingen, Germany). Detergents were removed by washing three times with buffer containing 8 M urea. The proteins were then reduced using DTT, alkylated using iodoacetamide, and the excess reagent was quenched by addition of DTT and washed through the filters. Buffer was exchanged by washing three times with 50 mM NH_4HCO_3 and proteins digested overnight by trypsin (Trypsin Gold, Promega, Madison, WI, USA) with an enzyme to protein ratio of 1:50. After overnight digestion, peptides were recovered by centrifugation and two additional washes using 50 mM NH_4HCO_3. Flowthroughs were combined, acidified, lyophilized and redissolved in 0.1% formic acid by sonication.

Nanoscale UPLC separation of tryptic peptides was performed with a nanoAcquity UPLC system (Waters Corporation, Manchester, UK) equipped with a HSS-T3 C18 1.8 µm, 75 µm × 250 mm analytical reversed-phase column (Waters Corporation, Manchester, UK) in direct injection mode (Tenzer *et al.*, 2011). Mobile phase A was water containing 0.1% v/v formic acid, while mobile phase B was ACN containing 0.1% v/v formic acid. Peptides were separated with a gradient of 5-40% mobile phase B at a flow rate of 300 nl/min. The analytical column temperature was maintained at 55 °C. The lock mass compound [Glu1]-Fibrinopeptide B (100 fmol/μl) was delivered by the auxiliary pump of the LC system at 500 nl/min to the reference sprayer of the NanoLockSpray source of the mass spectrometer.

Mass spectrometric analysis of tryptic peptides was performed using a Synapt G2-S HDMS mass spectrometer (Waters Corporation, Manchester, UK) equipped with a T-Wave-IMS device (Giles *et al.*, 2004) (Figure 1). All analyses were performed in positive mode ESI. The data were post-acquisition lock mass corrected using the doubly charged monoisotopic ion of [Glu1]-Fibrinopeptide B. Accurate mass LC-MS data were collected in DIA modes of analysis using MS^E in combination with online ion mobility separations. The spectral acquisition time in each mode was 0.6 s with a 0.05 s-interscan delay. In low energy MS mode, data were collected at constant CE of 4 eV. A CE ramp from 25 to 55 eV during each 0.6 s-integration was used as standard setting for the elevated energy MS scan in MS^E and HDMS^E mode. One cycle of low and elevated energy data was acquired every 1.3 s.

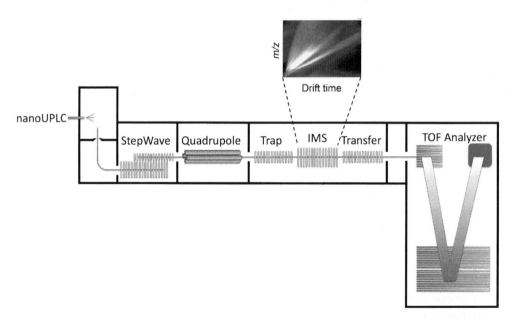

Figure 1. Schematic view of the Waters SynaptG2-S HDMS equipped with a traveling-wave ion mobility cell. After separation of the ions based on their ion mobilities in the IMS-cell using a traveling voltage wave, precursors are subjected to fragmentation in the transfer cell applying ramped collision energies.

Continuum LC-MS data were processed and searched using ProteinLynx Global SERVER version 3.0 (PLGS, Waters Corporation). Peptides had to meet the following search criteria: (1) trypsin as digestion enzyme; (2) up two missed cleavages allowed; (3) fixed carbamidomethylcysteine and variable methionine oxidation set as the modifications; (4) minimum length of six amino acids; (5) minimum three identified fragment ions. Only proteins identified by at least two peptides were considered. The false discovery rate (FDR) for peptide and protein identification was determined based on the search of a reversed database, which was generated automatically using PLGS 3.0 by reversing the sequence of each entry. The FDR was set to 1% for DB search in PLGS.

Post identification analysis including retention time alignment, EMRT (exact-mass-retention-time) and IMS clustering, normalization and label-free quantification was performed using the in-house developed software package ISOQuant.

Results and discussion

To compare classical MS^E-type of data acquisition with ion mobility enhanced MS^E (also referred to as $HDMS^E$), we analyzed 200 ng of tryptic cell lysates in 90 min gradient time. Using optimized mobility parameters to obtain maximal separation space in the mobility domain, we were able to identify over 2,500 proteins with the $HDMS^E$ approach, a 100% increase compared to MS^E without IMS. IMS provides an additional dimension of separation, improving overall system peak capacity while concomitantly reducing chimeric and composite interference. Furthermore, IMS increases selectivity of precursor-product ion alignment, as product ions can be aligned to their precursor ions using retention and drift times. However, as the inherent label-free quantification capabilities of the proprietary vendor software ProteinLynx GlobalSERVER (PLGS) are limited, we developed a software package, ISOQuant, which supports in-depth evaluation of DIA-based label-free IMS data. PLGS is used for raw data processing and for peptide and protein identification. ISOQuant automatically extracts LC-MS experiment data from PLGS and imports relevant information into a relational database (MySQL) subsequently applying a set of in-house developed and adapted third party analysis methods. Relations between multiple LC-MS runs are built and advanced statistics calculated. Non-linear retention time distortions between LC-MS runs are corrected (Podwojski *et al.*, 2009). Corresponding signals are clustered and subjected to multidimensional intensity normalization. Clusters are annotated by consensus peptides from associated LC-MS runs. Homologue proteins are filtered. Absolute in-sample amounts are calculated (Silva *et al.*, 2006). Finally, results of the performed analysis are exported as a set of uniform reports. ISOQuant provides easy access to routine application of label-free quantification by significantly reducing evaluation time and by offering standardized data evaluation procedures.

To demonstrate the performance of ISOQuant, we used data from tryptic cell lysates, which were separated by nanoUPLC and analyzed by LC-MS on a Waters Synapt G2-S mass spectrometer using 90 or 180 min gradients. Processing the data through the ISOQuant pipeline using cross-annotation of features increased the number of identified proteins compared to PLGS based results by more than 20% at a fixed protein FDR of 1%. Similar results were obtained on peptide level. Additionally, ISOQuant processing increased the number of proteins identified in 3/3 technical replicates from <50% to >97%, and also increased the number of peptides identified in all technical replicates about 80%.

In addition, high precision and accuracy of label-free quantification results could be demonstrated based on a hybrid proteome sample set with defined composition of tryptically digested proteomes from different species. Evaluation showed significant reduction of in-sample relative standard deviation compared to PLGS processed data and a significant increase in quantifiable proteins, reducing potential false-positive results. Additionally, in-depth analysis of the distribution of observed ratios between tested samples across all quantifiable proteins provides evidence for high quantification precision and accuracy of results obtained by ISOQuant processed data.

Acknowledgements

We thank all ISOQuant beta testers for their continuing critical evaluation of the software. We thank Ruben Spohrer for excellent sample preparation and H. Vissers and K. Richardson for discussions on data evaluation. This work was supported by Deutsche Forschungsgemeinschaft (INST 371/23-1 FUGG) to S.T., H.S., BMBF (e:Bio Express2Present, 0316179C) to S.T., the Forschungszentrum Immunologie (FZI), the Naturwissenschaftlich-Medizinische Forschungszentrum (NMFZ) and the Forschungszentrum Translationale Neurowissenschaften (FTN) of the Johannes Gutenberg University Mainz.

References

Giles, K., Pringle, S.D., Worthington, K.R., Little, D., Wildgoose, J.L. and Bateman, R.H., 2004. Applications of a travelling wave-based radio-frequency-only stacked ring ion guide. Rapid Commun Mass Spectrom 18: 2401-2414.

Podwojski, K., Fritsch, A., Chamrad, D.C., Paul, W., Sitek, B., Stuhler, K., Mutzel, P., Stephan, C., Meyer, H.E., Urfer, W., Ickstadt, K. and Rahnenfuhrer, J., 2009. Retention time alignment algorithms for LC/MS data must consider non-linear shifts. Bioinformatics 25: 758-764.

Silva, J.C., Gorenstein, M.V., Li, G.Z., Vissers, J.P. and Geromanos, S.J., 2006. Absolute quantification of proteins by LCMSE: a virtue of parallel MS acquisition. Mol Cell Proteomics 5: 144-156.

Tenzer, S., Docter, D., Rosfa, S., Wlodarski, A., Kuharev, J., Rekik, A., Knauer, S.K., Bantz, C., Nawroth, T., Bier, C., Sirirattanapan, J., Mann, W., Treuel, L., Zellner, R., Maskos, M., Schild, H. and Stauber, R.H., 2011. Nanoparticle size is a critical physicochemical determinant of the human blood plasma corona: a comprehensive quantitative proteomic analysis. ACS Nano 5: 7155-7167.

Wisniewski, J.R., Zougman, A., Nagaraj, N. and Mann, M., 2009. Universal sample preparation method for proteome analysis. Nat Methods 6: 359-362.

Small intestinal response to enterotoxigenic *Escherichia coli* infection in pigs as revealed by label free UPLC/MS^E proteomics

Laura Soler[1], Theo Niewold[1], Edwin de Pauw[2] and Gabriel Mazzucchelli[2]
[1]*Division of Livestock-Nutrition-Quality, Faculty of Bioscience Engineering, K.U. Leuven, Belgium; laura.solervasco@biw.kuleuven.be*
[2]*Laboratory of Mass Spectrometry-GIGA-Proteomics, University of Liège, Belgium*

Objectives

Enterotoxigenic *Escherichia coli* (ETEC) refers to non-invasive *E. coli* bacteria which adhere to the microvilli of small intestinal epithelial cells without inducing morphological lesions and producing enterotoxins that act locally on enterocytes (Schroyen *et al.*, 2012). Enteric diseases due to strains of ETEC are the most commonly occurring form of colibacillosis in pigs and man (Schroyen *et al.*, 2012). In neonatal and recently weaned piglets, ETEC-associated diarrhoea results in morbidity and mortality and is one of the economically most important diseases in swine husbandry (Schroyen *et al.*, 2012). The battle against this problem faces different (interrelated) obstacles: the understanding of the mechanisms involved in ETEC infection is relatively poor, no practical tools for the monitoring of the intestinal health status circumventing the inaccessibility of the intestinal system (biomarkers) are available, and there is a need for new preventive/therapeutic intervention schemes (Geens and Niewold, 2010). The high complexity of the intestinal system hampers the dissection of the biological effects and pathways involved in ETEC infection (Geens and Niewold, 2010). Functional genomics and transcriptomics have represented a partial solution to the latter, but still the true biochemical outcome of genetic information should be investigated through proteomics. However, proteomics is an emerging technique in animal sciences and its progress depends on increasing the sensitivity and throughput as well as the implementation of more specific and robust quantification methods (Bendixen *et al.*, 2011). In this study, we applied a simple, cost-effective, high-performance and sensitive approach to investigate the early pig proteomic intestinal response against ETEC infection for the first time.

Material and methods

The experimental animal protocol was approved by the local ethics committee for animal experiments in Leuven. Eight weaned piglets (Piètrain × Hypor) were divided in two groups and housed in two slatted floors with bedding, sufficiently separated pens, containing 4 piglets each. Pigs from each group received 5 ml PBS alone (mock), or 5 ml 109 cfu ETEC/ml PBS (infected) by force-feeding. This ETEC strain, *E. coli* O149K91, CVI-1000 (F4 (K88ac), LT +, STb +), was grown overnight in LB, centrifuged, and resuspended in PBS (pH 7.2) to an extinction at 600 nm, corresponding to 109 cfu/ml. At approximately 24 h after inoculation, the animals were euthanized by barbiturate overdose, weighed and sampled. A juxtaposed part (40 cm) of mid jejunum was cut out, flushed with 10 ml PBS (RT), opened, and approximately 30 cm of mucosa was scraped off with a glass slide, immediately frozen in liquid nitrogen, and stored at -80 °C. Approximately 200 mg of tissue was sonicated in lysis buffer (10 mM Tris-HCl pH 7.4, 4%SDS, 1% Proteinase Inhibitor cocktail (GE Lifesciences, USA) and 5 U/ml DNAse (Sigma-Aldrich, USA) and extracted proteins were collected in the supernatant after centrifugation. Protein content was determined with the RC/DC BioRad kit (USA) and concentration was adjusted at 6 µg/µl. Proteins were then reduced, alkylated, purified using a 2D cleanup kit (GE Lifesciences, USA) and trypsine-digested (Turtoi *et al.*, 2010). 3.5 µg of digested protein were dialyzed against 50% acetonitrile, 0.1% trifluoroacetic acid in water using C18 Zip-Tip devices (Millipore, USA). After SpeedVac concentration, tryptic peptides were resuspended to meet final conditions: 2.5 µg digested protein in 9 µl (final volume) 100 mM formiate buffer pH 10 containing 150 fmoles of MassPREP Digest Standards 1 (Waters, USA). Samples (in triplicates) were randomly analyzed using a nanoAcquity system (Waters, USA) coupled with the SYNAPT mass spectrometer (Waters, USA). 9 µl of sample was injected, corresponding to a sample load of 2.5 µg. This approach combines separation using pH 10 in the first and pH 2.6 in the second separation dimension using reverse phase columns. The LC system configuration was as follows: trap column Symmetry C18 5 µm, 180 µm × 20 mm (Waters, USA), analytical column BEH C18 1.7 µm, 75 µm × 250 mm (Waters, USA), solvent A (0.1% formic acid in water), solvent B (0.1% formic acid in acetonitrile) and flow rate 300 nl/min (gradient: 0 min, 97% A; 90 min, 60% A). The following acquisition parameters were set for the SYNAPT mass spectrometer: data independent, alternate scanning (MS^E) mode, 50-1,990 *m/z* range, ESI+, V optics, scan time 1 s, cone 24 V and lock mass [Glu1]-Fibrinopeptide B $[M+2H]^{2+}$ 785.8426 *m/z*. LC-MS data were processed and searched using ProteinLynx GlobalServer version 2.3 (PLGS 2.3) (Waters, USA). Raw data sets were processed including ion detection, deisotoping, deconvolution, and peak lists generated based on the assignment of precursor ions and fragments based on similar retention times. The principles of the applied data clustering and normalization have been explained elsewhere (Silva *et al.*, 2006). The database search algorithm was described elsewhere (Li *et al.*, 2009). The Swiss-ProtKB/ Swiss-Prot *Sus scrofa* database with proteins of MassPREP Digest Standards 1 (Waters, USA) appended as internal standards was employed. The protein identifications were based on the detection of at least 3 fragment ions per peptide 7 fragments per protein, 1 peptide per protein in all the technical replicates. A maximum false positive rate of 4% was allowed. The search tolerances were automatically read from the data and were approximately 9 and 23 ppm for the precursor and product ions, respectively. Peptide modifications carbamidomethylation and oxidation (M) were set as variable. One missed cleavage for trypsin digestion was allowed. A manual data filter for relative comparison purposes was included, additional to the default software intelligent scoring filter, specifying that only proteins identified in all the biological replicates of each group (control vs. ETEC) would be included in the comparison analysis. The transcriptomic response towards ETEC as determined by microarray and qPCR in these animals in previous studies (Niewold *et al.*, 2012) was compared with the proteomic response.

Results and discussion

After PLGS autonormalization, approximately 1,680±280 proteins were accurately identified in each of the studied samples. A total of 1,560 proteins were detected in all the biological replicates of the studied groups and subsequently included in the comparative expression analysis. The significance of regulation level was specified at 30%. Hence, 1.3-fold (±0.30 natural log scale) change was used as a threshold to identify significantly up- or down-regulated expression; this is typically 2-3 times the estimated error of the intensity measurement. All protein hits that were identified with a confidence of >95% were regarded as genuine up- or down-regulated proteins. 122 proteins were confidently found to be down-regulated in ETEC-infected pigs while 99 proteins were up-regulated. Two proteins were identified only in ETEC samples and 2 proteins only in control samples. Several punctual disagreements were found between the transcriptomic and proteomic expression results. Interestingly, we identified a strong overexpression of different proteins involved in post-transcriptional regulation, including PCBP3, PCBP2, TCEA1 and PSMA, which would explain those incongruities. However, several hypothesis suggested by transcriptomics but yet unresolved were confirmed. A simpler model of ETEC-driven host reaction was established by a single ETEC challenge not enough to cause infection (confirmed by a total jejunal clearance of ETEC after 24 h), thus evaluating effects beyond the acute reaction to the inoculums (Niewold *et al.*, 2012). The high performance proteomic analysis performed in this study allowed us to describe the pathways involved in the innate immune responses developed in the pig intestinal response in this model, whose interpretation was not straight forward. In brief, an activation of the innate immune response at the expense of the adaptive immune response was observed, including the overexpression of several innate immunity-related proteins such as haptoglobin, Hsp70, PigMAP, Serpin 3, HMGP, MIF and fibrinogen among others. The depletion of several antimicrobial secreted defensins from epithelial cells including prophenins, protegrins and cathelins was observed as an effect of ETEC exposure, in agreement with previous transcriptional results (Niewold *et al.*, 2012). Interestingly, another group of antibacterial intra-epithelial proteins (Histone-like H1 proteins) was strongly up regulated (20 to 60-fold times) in ETEC-challenged pigs, indicating a defence mechanism not well explored in pigs. In conclusion, the label free UPLC/MS^E proteomic approach applied in this study was confirmed as a valuable tool to dissect of the events involved in the pig innate immune response despite the high complexity of the intestinal system.

Acknowledgements

This study was financed by the Martín Alonso Escudero Foundation.

References

Bendixen, E., Danielsen, M., Hollung, K., Gianazza, E. and Miller, I., 2011. Farm animal proteomics--a review. J Proteomics 74: 282-293.

Geens, M.M. and Niewold, T.A., 2010. Preliminary Characterization of the Transcriptional Response of the Porcine Intestinal Cell Line IPEC-J2 to Enterotoxigenic Escherichia coli, Escherichia coli, and E. coli Lipopolysaccharide. Comp Funct Genomics 2010: 469583.

Li, G.Z., Vissers, J.P., Silva, J.C., Golick, D., Gorenstein, M.V. and Geromanos, S.J., 2009. Database searching and accounting of multiplexed precursor and product ion spectra from the data independent analysis of simple and complex peptide mixtures. Proteomics 9: 1696-1719.

Niewold, T.A., Schroyen, M., Geens, M.M., Verhelst, R.S.B. and Courtin, C.M., 2012. Dietary inclusion of arabinoxylan oligosaccharides (AXOS) down regulates mucosal responses to a bacterial challenge in a piglet model. Journal of Functional Foods 4: 626-635.

Schroyen, M., Stinckens, A., Verhelst, R., Geens, M., Cox, E., Niewold, T. and Buys, N., 2012. Susceptibility of piglets to enterotoxigenic Escherichia coli is not related to the expression of MUC13 and MUC20. Animal Genetics 43: 324-327.

Silva, J.C., Gorenstein, M.V., Li, G.Z., Vissers, J.P.C. and Geromanos, S.J., 2006. Absolute quantification of proteins by LCMSE – A virtue of parallel MS acquisition. Molecular & Cellular Proteomics 5: 144-156.

Turtoi, A., Mazzucchelli, G.D. and De Pauw, E., 2010. Isotope coded protein label quantification of serum proteins-Comparison with the label-free LC-MS and validation using the MRM approach. Talanta 80: 1487-1495.

Rapid protein production of Atlantic Salmon, *Salmo salar*, serum amyloid A (SAA) in an inducible *Leishmania tarentolae* expression system

Mark Braceland[1], Mark McLaughlin[1], P. David Eckersall[1] and Mangesh R. Bhide[2]
[1]*Institute of Infection, Immunity and Inflammation, University of Glasgow, Bearsden Rd Glasgow, United Kingdom; m.braceland.1@research.gla.ac.uk*
[2]*Laboratory of Biomedical Microbiology and Immunology, Department of microbiology and immunology, University of Veterinary Medicine and Pharmacy, Kosice, Slovakia*

Introduction

Atlantic salmon, *Salmo salar*, farming is one of the most economically important industries within aquaculture. The production of this livestock has, in recent times, exhibited an unprecedented growth, in terms of food production; rising from an estimated 13,300 tonnes in 1982 to 1,5 million in 2011 and in light of current global population growth is projected to keep growing. However, in order for growth to continue in such a fashion a number of obstacles must be overcome. Whilst these include a number of logistical issues arguably from an ecological and economic view one of the largest problems salmon production, and indeed aquaculture as a whole, faces is that of disease.

Infection of Atlantic salmon, *Salmo salar*, and rainbow trout, *Oncorhynchus mykiss*, in Norway with salmon alpha virus (SAV) has marked economic impact on their cultivation and leads to pancreatic disease (PD). This atypical alphavirus is transmitted horizontally causing a significant economic impact on the aquaculture industry. A number of diagnostic tools for detecting PD are widely used, for instance, histopathology and virus identification in tissues by real time polymerase chain reaction (RT-PCR SAV). These methods are both invasive and destructive as tissue samples are taken post-mortem, further adding to mortality on an aquaculture. Therefore using a non-destructive method such as pathogen-specific antibody detection in serum may be advantageous. However, it has been shown that fish immunoglobulin (Ig) is synthesized much slower than in mammals and in temperate species such as salmon can take up to six weeks to be synthesized. Therefore, interest has developed in assessment of how the humoral components of the innate immune system respond to PD with a potential they may serve as non-destructive biomarkers of disease pathology in fish.

Serum Amyloid A (SAA) is an acute phase protein (APP) involved in both inflammatory and humoral anti-infective responses. This protein is well conserved throughout vertebrates and has been shown in a number of genomic studies to be up regulated during a number of diseases in various teleost species (Bayne and Gerwick, 2001). For example, a 500 fold induction of A-SAA has been shown in zebra fish infected with bacteria species *Aeromonas salmonicida* or *Staphylococcus aureus* (Lin *et al.*, 2007) and is also up-regulated during viral infection in the salmonid rainbow trout, *Oncorhynchus mykiss* (Rebl *et al.*, 2009). Moreover, LPS treatment has shown to stimulate SAA5 (NM_001146565.1) expression in salmon hepatocytes (Jorgensen *et al.*, 2000). Expression was also increased by 1.1, 9.1 and 27.5 fold at 1, 3 and 5 days post infection respectively, in the head kidney of salmon infected with SAV (Herath *et al.*). Despite this genomic evidence that SAA is an APP in salmon, a salmon specific antibody or cross-reacting with salmon SAA has not been reported. This is needed to quantify levels of SAA in the serum during disease. Thus, the concentration of this protein in the serum under homeostasis and how this alters during disease is not known. Therefore, salmon SAA was the chosen candidate for *in vivo* recombinant production in a *Leishmania* system to provide antigen for antibody production in order to develop a specific ELISA.

Materials and methods

Total RNA was isolated from the liver of an Atlantic salmon using the reagent Purezol (Bio-Rad) in accordance with manufacturer's instructions. cDNA was then synthesised by reverse transcription of 750 µg of RNA following protocol provided with Revert Aid (Thermo Scientific). Amplification of SAA5 was carried out using oligonucleotide primers (Table 1) designed using SAA5 sequence NM_001146565.1. PCR product was ran at 70 V on a 2% agarose (+ ethidium bromide) gel to check product was ran to the correct estimated mass.

These primers (Table 1) contained linkers (or overlapping regions) so that the PCR product from amplification of SAA5 could be incorporated, by OE-PCR, into the expression cassette. The cassette constructed, following Jenna Bioscience instructions, consisted of (from 5' to 3') end: flanking region for incorporation into chromosome at ODC region by double crossover, T7 promoter, signal peptide, overlap region for OE-PCR, salmon SAA5 gene, overlap region, factor FxA site, GFP tag, Myc tag, stop codon, UTR, gene for resistance against bleomycin, and the 3' end flanking region.

L. tarentolae was used for the production of recombinant Atlantic salmon SAA (Sugino and Niimi). In brief, cassette was incorporated into *Leishmania* by electroporation (at 450 V) post mixing of linear DNA with the vector, with double crossover taking place. The population was then allowed to grow for 24 hours before the addition of bleomycin, which selects for cells in which electroporation has been successful. After subsequent passaging and growth of

Table 1. Primers used to amplify Atlantic salmon serum amyloid A.

SAA reverse
GCTTCTCCCTTCTATGGTACCCTTAAGGTAGTTCCTTGGGAGTCCATT
SAA forward
GCTGGCGCCTCTCTAGACACACCTGGTGAAGCTGCTCGAGGT

population conserves were then made at a 3:1 ratio with BHI medium (containing the cells) and 85% glycerol. One conserve of 1.6 ml was then reactivated in 10 ml of BHI medium and left until optical density was 1.4 at which point tetracycline was added to induce production of protein. Production was visualised via fluorescent microscopy as the GFP tag used in the expression cassette is only expressed after tetracycline treatment.

Results and discussion

The PCR product from amplification of cDNA was estimated to be just under 300 bp long after gel checking (Figure 1). The product was then sequenced and was found to be same length as the estimated size of 293 bp (Figure 1) and share 100% homology with Atlantic salmon SAA5 sequence NM_001146565.1.

Post tetracycline induction expression of the GFP tag was observed via green fluorescence under ultraviolet (UV) light condition. From this we can infer the successful expression of the expression cassette and thus SAA. Although expression of the GFP tag linked to SAA has been successful it is now necessary to scale up the production in order to purify enough protein to raise sufficient antibody in order to develop a specific salmon SAA ELISA. Purification will be from medium (as protein is excreted from cells) using either Myc or GFP affinity beads. Once protein is bound Fxa protease will be used to cleave SAA from the linked tags allowing production of antibody and ELISA development. This is a major step toward evaluating the diagnostic value of monitoring this APP in the serum of salmon during PD and other diseases

Conclusion

This project has successfully transfected salmon SAA5 into *Leishmania tarentolae*, producing conserves which will be used to scale up the expression of SAA in order to attain enough protein for antibody production and subsequent ELISA development.

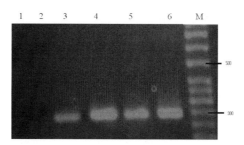

Figure 1. PCR product. 2% Agarose gel (+ ethidium bromide) showing the size of PCR product after PCR using designed primers of salmon SAA. Lanes 1 to 4 all contain PCR product, whilst lane possesses a master mix which underwent PCR conditions but possessed no salmon cDNA and lane 6 contains loading dye and water only.

Acknowledgements

Dr. Mangesh Bhide and his colleagues at the University of Veterinary Medicine and Pharmacy in Kosice are gratefully thanked for their constant guidance and friendship during the STSM and COST-Farm Animal Proteomics for allowing this STSM to take place.

References

Bayne, C.J. and Gerwick, L., 2001. The acute phase response and innate immunity of fish. Dev Comp Immunol 25: 725-743.

Herath, T.K., Bron, J.E., Thompson, K.D., Taggart, J.B., Adams, A., Ireland, J.H. and Richards, R.H., Transcriptomic analysis of the host response to early stage salmonid alphavirus (SAV-1) infection in Atlantic salmon Salmo salar L. Fish Shellfish Immunol 32: 796-807.

Jorgensen, J.B., Lunde, H., Jensen, L., Whitehead, A.S. and Robertsen, B., 2000. Serum amyloid A transcription in Atlantic salmon (Salmo salar L.) hepatocytes is enhanced by stimulation with macrophage factors, recombinant human IL-1 beta, IL-6 and TNF alpha or bacterial lipopolysaccharide. Dev Comp Immunol 24: 553-563.

Lin, B., Chen, S., Cao, Z., Lin, Y., Mo, D., Zhang, H., Gu, J., Dong, M., Liu, Z. and Xu, A., 2007. Acute phase response in zebrafish upon Aeromonas salmonicida and Staphylococcus aureus infection: striking similarities and obvious differences with mammals. Mol Immunol 44: 295-301.

Rebl, A., Goldammer, T., Fischer, U., Kollner, B. and Seyfert, H.M., 2009. Characterization of two key molecules of teleost innate immunity from rainbow trout (Oncorhynchus mykiss): MyD88 and SAA. Vet Immunol Immunopathol 131: 122-126.

Sugino, M. and Niimi, T., Expression of multisubunit proteins in Leishmania tarentolae. Methods Mol Biol 824: 317-325.

Proteomic profiling of cerebrospinal fluid in canine degenerative myelopathy

Viviana Greco[1,2*], Intan Nur Fatiha[3*], Jacques Penderis[3], Paul Montaque[3], Mark McLaughlin[3*], Andrea Urbani[1,2] and Thomas J. Anderson[3]

[1]Proteomic and Metabonomic Laboratory, Fondazione Santa Lucia, Rome, Italy; vivianagreco82@yahoo.it

[2]Department of Experimental Medicine and Surgery, University of Rome 'Tor Vergata', Rome, Italy

[3]School of Veterinary Medicine, College of Medicine, Veterinary Medicine and Life Science, University of Glasgow, Glasgow, United Kingdom

*these authors equally contribueted to the study

Objectives

Canine *degenerative myelopathy* (DM) is an incurable, progressive disease of the canine spinal cord that has similarities with human amyotrophic lateral sclerosis (ALS). Onset is typically after the age of 7 years and was reported initially to occur frequently in the German shepherd dog, although recent reports on DM in Pembroke Welsh corgi, and boxer dog has also emerged (Coates and Wininger, 2010). Progressive weakness and incoordination of the rear limbs are often the first signs seen in affected dogs, with progression over time to complete paralysis with elective euthanasia the outcome (Johnston *et al.*, 2000).

Although the pathological pathways and underlying DM is not yet fully known, a recent finding that a mutation in the Sod1 gene is associated with DM has strengthened the proposal that DM is similar to fALS which is also frequently associated with mutations in SOD1 (Awano *et al.*, 2009). However, genetic analysis has found that some cases homozygous for the Sod1 mutation fail to develop the clinical signs and conversely cases that present the clinical features of DM do not harbour the Sod1 mutation. While the combination of the clinical evaluation of DM with the genetic analysis for the presence of the Sod1 mutation may strengthen the diagnosis, they are not definitive. Therefore it is necessary to identify clinical biomarkers. Cerebrospinal fluid (CSF) is the only body fluid that surrounds the brain, and the analysis of CSF can provide direct information regarding the physiological condition of the brain and spinal cord (Di Terlizzi and Platt, 2009). The aim of this study is to perform a proteomic analysis of CSF of dogs to establish that our protocols on human CSF are compatible with this canine CSF. We have also proposed an assessment of the peptide profile of CSF from cases diagnosed with degenerative myelopathy to determine if the characteristic CSF protein patter is comparable with dogs with idiopathic epilepsy (IE) as control group.

Material and methods

Clinical CSF material for this study was obtained from the neurological service, University of Glasgow, Small Animal Hospital with appropriate ethical permission. All cases were subjected to a rigorous neurological examination. Sampling was performed in conjunction with myelography or MRI procedures. CSF was obtained from cerebellomedullary cistern or lumbar subarachnoid space during anaesthesia. Radiographic, MRI and EMG examinations were conducted under general anaesthesia. Routine haematology and biochemistry evaluation was conducted in each patient to evaluate general health and assess for concurrent systemic diseases. DNA was extracted from blood samples and the genetic status of the Sod1 gene assesed using RFLP based on the loss of a HypAV digestions site due to the presence of the 118G>A point mutation. All DM cases were homozygous for this mutation.

Samples were extracted and analyzed according to already reported method for human CSF (Del Boccio *et al.*, 2007). For C4 ZipTip extraction, the solid-phase material was first activated by multiple washing with 10 µl of ACN/water (1:1) and then equilibrated by 0.1% TFA. Briefly, CSF samples were first acidified and CSF proteins were extracted by ZipTip C4 tips (Millipore) with a sandwich layer method on MTP Ground steel 384 (Bruker Daltonics, Germany). For analyzing peptides and proteins, a sinapinic acid matrix seed layer was firstly created by depositing a droplet of a saturated solution of sinapinic acid in 100% ethanol on the target. Thereafter, 2 µl of the matrix dissolved in 30% ACN and 0.1% TFA was used to elute the sample from the resin. Advantages of this thin layer method are that it provides a greater tolerance to impurities such as salts and detergents, a better resolution, and an higher spatial uniformity. This method is especially useful for the accurate mass determination of proteins.

All analysis were performed with a Ultraflex III MALDI-TOF mass spectrometer (Bruker Daltonics, Germany) in the mass range 2-20 kDa. Instrument calibration was performed using Proteins Standard I (Bruker Daltonics, Germany) as external calibrators. Acquired mass spectra were processed using Bruker software flexAnalysis 3.0.

Proteomic analysis is supported by technologies such as Ultra Performance Liquid Chromatography (UPLC) that is promising high-throughput approaches to identify new potential biomarkers in various body fluids.

Results and discussion

In order to obtain high-resolution linear MALDI-TOF-MS spectra of the investigated samples we initially developed an optimized method for sample preparation. Under the conditions described in materials and methods, the highest mass resolution and the lowest background level was obtained, and the 2-20 kDa mass range was explored in order to investigate the low protein molecular weight region in the CSF.

By this investigation, a differential protein profiling among Myelopathy *vs* Epilepsy groups was highlighted. These differences are evident in low mass range (2,000-4,000 m/z) and higher mass range (9,000-15,000 m/z) (Figure 1).

Particularly, e.g. the peak m/z 13,181 was characteristic of DM (a); e.g. the peak m/z 14,088 was characteristic of IE (b), whereas the peak m/z 11,515 was present in both groups (c).

These preliminary results suggest changes in protein levels (e.g. transthyretin, TTR) related to the pathogenic mechanism of this disease.

These preliminary results suggest changes in protein levels related to the pathogenic mechanism of this disease and may distinguish DM from idiopathic epilepsy that is not associated with neurodegenerative processes. It remains to be determined if this analysis can differential DM from cases of disc herniation that present similar clinical symptoms and often leads to misdiagnosis.

Also in humans the analysis of the CSF proteome has been used to search for diagnostic markers in patients with neurological disease. DM is a neurological disease of dogs similar in many ways to human amyotrophic lateral sclerosis (ALS) and strengthened by the involvement of a SOD1 mutation in both diseases. To date there are no naturally occurring large animal models of ALS and research into ALS has relied heavily on transgenic animal models.

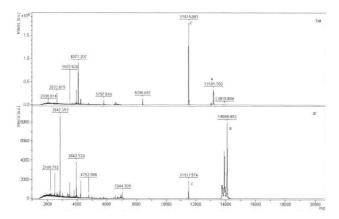

Figure 1. Comparison of average MALDI TOF mass spectra for dogs with degenerative myelopathy (DM) and idiopathic epilepsy (IE). CSF average mass spectrum of DM and average spectrum of IE group, respectively, were reported. The letter A is one of the characteristic peaks of the DM; B is one of the characteristic peaks of the IE and C denotes one of the peaks common to the two clinical groups.

For this reason we have pursued a comparative investigation between the CSF of dogs with DM and CSF of ALS patients. In fact we speculate that such a comparative analysis may highlight molecular features common to both pathologies and may provide additional information about these diseases.

References

Awano, T., Johnson, G.S., Wade, C.M., Katz, M.L., Johnson, G.C., Taylor, J.F., Perloski, M., Biagi, T., Baranowska, I., Long, S., March, P.A., Olby, N.J., Shelton, G.D., Khan, S., O'Brien, D.P., Lindblad-Toh, K. and Coates, J.R., 2009. Genome-wide association analysis reveals a SOD1 mutation in canine degenerative myelopathy that resembles amyotrophic lateral sclerosis. Proc Natl Acad Sci U S A 106: 2794-2799.

Coates, J.R. and Wininger, F.A., 2010. Canine degenerative myelopathy. Vet Clin North Am Small Anim Pract 40: 929-950.

Del Boccio, P., Pieragostino, D., Lugaresi, A., Di Ioia, M., Pavone, B., Travaglini, D., D'Aguanno, S., Bernardini, S., Sacchetta, P., Federici, G., Di Ilio, C., Gambi, D. and Urbani, A., 2007. Cleavage of cystatin C is not associated with multiple sclerosis. Ann Neurol 62: 201-204; discussion 205.

Di Terlizzi, R. and Platt, S.R., 2009. The function, composition and analysis of cerebrospinal fluid in companion animals: part II – analysis. Vet J 180: 15-32.

Johnston, P.E., Barrie, J.A., McCulloch, M.C., Anderson, T.J. and Griffiths, I.R., 2000. Central nervous system pathology in 25 dogs with chronic degenerative radiculomyelopathy. Vet Rec 146: 629-633.

Automatic prediction of PTMs in *Ehrlichia ruminantium* – creating new datasets for Quickmod analyses

Miguel Ventosa[1,2], Oliver Horlacher[3], Nathalie Vachiéry[4], Thierry Lefrançois[5], Ana V. Coelho[2], Frederique Lisacek[3] and Isabel Marcelino[1,2]

[1]*IBET-Instituto de Biotecnologia Experimental e Tecnológica; Apartado 12, 2780-901 Oeiras, Portugal*

[2]*ITQB-Instituto de Tecnologia Química e Biológica, Universidade Nova de Lisboa; Av. da República, 2780-157 Oeiras, Portugal; miguelv@itqb.unl.pt*

[3]*SIB – Swiss Institute of Bioinformatics, University Medical Center; 1, rue Michel Servet, 1211 Geneva 4, Switzerland*

[4]*Centre de coopération Internationale en Recherche Agronomique pour le Développement, UMR CMAEE, 97170 Petit-Bourg, Guadeloupe, FWI*

[5]*Centre de coopération Internationale en Recherche Agronomique pour le Développement, UMR CMAEE, UMR CMAEE, 34398 Montpellier, France*

Objectives

Ehrlichia ruminantium (ER) is an obligate intracellular bacterium, from the order *Rickettsiales*, which causes Heartwater, a fatal tick-borne disease in ruminants. This disease is a major limitation to livestock production in sub-Saharan Africa and in some Caribbean islands. Recent studies showed that key proteins such as the Major Antigenic Protein 1 (MAP1) are glycosylated (Postigo *et al.*, 2008) and that about 25% of ER proteome account for isoforms, indicating the importance of post-translational modifications (PTMs) in the *ER* infection process (Marcelino *et al.*, 2012).

The role of bioinformatics in the analysis of mass spectrometry data has become essential mainly due to the size of datasets this technique can produce. To process data efficiently, new software packages and algorithms are continuously being developed in order to improve protein identification and characterization in terms of high-throughput and statistical accuracy (Kumar and Mann, 2009). One of the tools for data analysis is Quickmod (Ahrne *et al.*, 2011a), a software that assists with the process of PTM detection.

Herein, we focus on the automatic prediction of PTMs on ER protein samples. For this, it was necessary to improve the analysis of mass spectrometry data and create an easy methodology for the bioinformatics discovery of PTMs.

Material and methods

E. ruminantium Gardel (ERG) strain isolated in Guadeloupe (FWI) was used throughout these studies. ERG was routinely cultivated *in vitro* in bovine aortic endothelial cells, BAE (Marcelino *et al.*, 2005; Pruneau *et al.*, 2012). Infectious, extracellular elementary bodies (EBs) were harvested at 80% cell lysis and purified by a multistep centrifugation process (Marcelino *et al.*, 2007). Total ER protein extracts were obtained using sonication, quantified by 2D Quant kit (GE, Uppsala, Sweden) and analyzed using 2DE gels, pH 3-10. All the protein spots were digested with trypsin and analysed by a MALDI-TOF/TOF 4800 *plus* equipment.

For each protein spot, a maximum of twelve most intense peaks was submitted to fragmentation. Each spectrum was identified using the MASCOT search engine (Perkins *et al.*, 1999) and results were saved as pep.xml files. Prior to processing this data with Quickmod, similar peptides were clustered and a spectral library was created. A visualization tool was developed for assessing the validity of clusters. This allowed us to evaluate the performance of the clustering algorithm and the quality of the spectra. The spectral library was then constructed using the Liberator and Deliberator tools (SIB) (Ahrne *et al.*, 2011b)*(Ahrné, Ohta, et al.* 2011). These tools enable the user to create custom spectral libraries from peptide identification data in the pep.xml format and experimental data in the mgf format.

Results and discussion

An extensive number of *ER* protein sequences are freely available. Our searches with MASCOT used small database sets from UniprotKB and Uniref in order maximise the number of spectra identified.

The developed visualization tool facilitated the evaluation of the clustering process and led to two main results. By combining the pep.xml files generated by MASCOT with the corresponding MS/MS spectra, we confirmed that the peptide sequences concur with the majority of clusters. The spectra were grouped in a total of 122 clusters, two of which are illustrated in Figure 1.1a and 1.1b.

Figure 1.1.a represents a nonagon cluster that includes the outcome of the digestion of the protein spots marked in zoom 1.1., and identified as MAP1. The MAP1 protein (from Major Antigenic Protein) is encoded by a immunodominant polymorphic gene (Allsopp *et al.* 2001) and MAP1-family proteins are considered as priority targets for candidates vaccines (Frutos *et al.*, 2006; Marcelino *et al.*, 2012). This clustering approach allows the easy detection of proteins with several different PTMs. It can also be used for smaller clusters with two, three and four spectra as shown in Figure 1.2b for the chaperone protein HtpG.

Furthermore, this analysis allowed us to distinguish MS/MS peptide spectra of sample contaminants/artifacts. These spectra are grouped in huge clusters. We believe that these

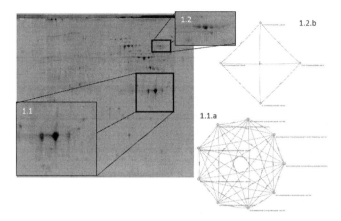

Figure 1. Example of output of the visualization tool. Small changes in the isoelectric point and in the molecular of the protein lead to different spots in the gel. With the clustering and its visualization it is very easy to point the protein spots that we should focus our attention. Figure 1.1a includes the peptides from the protein annotated as MAP 1 (Major antigenic protein 1) while Figure 1.2 includes the peptides from the protein annotated as Chaperone HtpG.

could be due to the alpha-cyano-4-hydroxycinnamic acid MALDI matrix used for MS data acquisition or from polymers, due to the destaining of the gel spots in multi-well plates.

After this thorough analysis of our MS/MS data obtained with MALDI-TOF-TOF and the establishment of a suitable spectra library for ER peptides, Quickmod can be run. In fact, spectra that were not identified with the database search engine, can be scrutinized by Quickmod. This algorithm searches for mass deviation between input spectra and consensus spectra representing each cluster in the spectral library previously built (on-going work).

In summary, the use of MS/MS data obtained by MALDI-TOF-TOF led to (1) optimize spectral libraries and thereby (2) enhance the data to be further processed in Quickmod analyses. This work also allowed establishing a new collaboration between IBET/ITQB and SIB for the characterization of PTM in ER, and possibly in other Rickettsiales pathogens with impact in human and animal health.

Acknowledgements

Authors acknowledge funding from project ER-TRANSPROT (PTDC/CVT/114118/2009) as well as a Post-doc grant SFRH/ BPD/ 45978/ 2008 (I. Marcelino) all by Fundação para a Ciência e a Tecnologia (Lisboa, Portugal); as well as EU project FEDER 2007-2013, FED 1/1.4-30305 and COST action FA-1002 Proteomics in Farm Animals for networking opportunities.

References

Ahrne, E., Nikitin, F., Lisacek, F. and Muller, M., 2011a. QuickMod: A tool for open modification spectrum library searches. J Proteome Res 10: 2913-2921.

Ahrne, E., Ohta, Y., Nikitin, F., Scherl, A., Lisacek, F. and Muller, M., 2011b. An improved method for the construction of decoy peptide MS/MS spectra suitable for the accurate estimation of false discovery rates. Proteomics 11: 4085-4095.

Frutos, R., Viari, A., Ferraz, C., Morgat, A., Eychenie, S., Kandassamy, Y., Chantal, I., Bensaid, A., Coissac, E., Vachiery, N., Demaille, J. and Martinez, D., 2006. Comparative Genomic Analysis of Three Strains of Ehrlichia ruminantium Reveals an Active Process of Genome Size Plasticity. J Bacteriol 188: 2533-2542.

Kumar, C. and Mann, M., 2009. Bioinformatics analysis of mass spectrometry-based proteomics data sets, FEBS Lett, Netherlands, pp. 1703-1712.

Marcelino, I., de Almeida, A.M., Brito, C., Meyer, D.F., Barreto, M., Sheikboudou, C., Franco, C.F., Martinez, D., Lefrancois, T., Vachiery, N., Carrondo, M.J., Coelho, A.V. and Alves, P.M., 2012. Proteomic analyses of Ehrlichia ruminantium highlight differential expression of MAP1-family proteins. Vet Microbiol 156: 305-314.

Marcelino, I., Vachiery, N., Amaral, A.I., Roldao, A., Lefrancois, T., Carrondo, M.J., Alves, P.M. and Martinez, D., 2007. Effect of the purification process and the storage conditions on the efficacy of an inactivated vaccine against heartwater. Vaccine 25: 4903-4913.

Marcelino, I., Verissimo, C., Sousa, M.F., Carrondo, M.J. and Alves, P.M., 2005. Characterization of Ehrlichia ruminantium replication and release kinetics in endothelial cell cultures. Vet Microbiol 110: 87-96.

Perkins, D.N., Pappin, D.J., Creasy, D.M. and Cottrell, J.S., 1999. Probability-based protein identification by searching sequence databases using mass spectrometry data, Electrophoresis, Germany, pp. 3551-3567.

Postigo, M., Taoufik, A., Bell-Sakyi, L., Bekker, C.P., de Vries, E., Morrison, W.I. and Jongejan, F., 2008. Host cell-specific protein expression *in vitro* in Ehrlichia ruminantium. Vet Microbiol 128: 136-147.

Pruneau, L., Emboule, L., Gely, P., Marcelino, I., Mari, B., Pinarello, V., Sheikboudou, C., Martinez, D., Daigle, F., Lefrancois, T., Meyer, D.F. and Vachiery, N., 2012. Global gene expression profiling of Ehrlichia ruminantium at different stages of development. FEMS Immunol Med Microbiol 64: 66-73.

Tandem mass spectrometry for species recognition and phenotyping in fish

Tune Wulff[1,2], Flemming Jessen[1], Magnus Palmblad[2] and Michael Engelbrecht Nielsen[1]
[1]Biological Quality Group, National Food Institute, Technical University of Denmark, Kgs. Lyngby, Denmark; tuwu@food.dtu.dk
[2]Biomolecular Mass Spectrometry Unit, Department of Parasitology, Leiden University Medical Center, Leiden, the Netherlands

Objectives

Authentication of food products is a major concern throughout the food industry. From husbandry as well as aquaculture, even in mildly processed products it is often impossible to visually identify the origin, making the use of molecular methods necessary for species validation. Within the field of proteomics, different methods have been implemented, either by taking advantage of specific mass spectrometry profiles (Mazzeo *et al.*, 2008) or using specific peptides or proteins as means to correctly identify the origin of the sample (Berrini *et al.*, 2006; Carrera *et al.*, 2007). We present here an unguided and robust new proteomics-based method for species recognition and product authentication, exemplified by a study of muscle samples from 17 different fish species. Based on large-scale pairwise comparison of tandem mass spectra, similarities and differences in the proteome are used to differentiate between the species. In addition we studied how different tissues from the same animal, commercially important rainbow trout (*Oncorhynchus mykiss*), have clearly distinguishable proteome profiles as measured by tandem mass spectral content.

Material and methods

Muscle tissue from 17 different fish species were provided from the EU project 'Advanced methods for identification and quality monitoring of (heat) processed fish' (FAIR CT95 1227).

A total of five adult rainbow trout individuals were euthanized using an overdose of MS222 after which the different organs/tissues were isolated and sampled. The following seven different tissues/organs were dissected: spleen, kidney, blood cells, plasma, heart, skeletal muscle and eye lens. Immediately after dissection, tissue samples were transferred to a cryotube and snap frozen in liquid nitrogen followed by storage at -80 °C until protein extraction. During dissection, the eye was taken as a whole and the eye lens was removed at day of sample preparation while the eye was still frozen. The blood samples were kept on ice until centrifugation at 5,000×g at 4 °C for 5 min after which the supernatant (plasma) and the pellet (blood cells) was collected and stored at -80 °C. Proteins were extracted in 100 µl Urea buffer [8 M Urea, 40 mM MgCl, and 50 U/ml Benzonase®] with 0.5 mm zirconium oxide beads, by homogenization using an air-cooled Bullet Blender® (Next Advance Inc.). The procedure involved 3 min homogenization followed by incubation for 12 min at 4 °C then 1 min homogenization and incubation for 12 min at 4 °C. Finally samples were centrifuged for 30 min at 4 °C after which the supernatants were collected. The final protein concentration was measured using a bicinchoninic acid (BCA) protein assay kit (Thermo Fischer Scientific).

Liquid chromatography-tandem mass spectrometry

The in-solution digests of the organ extracts were separated using splitless NanoLC-Ultra 2D plus (Eksigent, Dublin, CA) parallel ultra-high pressure liquid chromatography (UHPLC) systems with an additional loading pump for fast sample loading and desalting. Each UHPLC system was configured with two 300 μm-i.d. 5-mm PepMap C18 trap columns (Thermo Fischer Scientific) and two 15-cm 300 μm-i.d. ChromXP C18 columns (Eksigent). All samples were separated by 45-minute linear gradients from 4 to 33% acetonitrile in 0.05% formic acid. The UHPLC systems were coupled on-line to amaZon ETD speed high-capacity 3D ion traps with standard ESI sources (Bruker Daltonics, Bremen, Germany). After each MS scan, up to ten abundant multiply charged species in the m/z 300-1,300 range were automatically selected for MS/MS but excluded for one minute after being selected twice. The UHPLC systems were controlled using HyStar 3.4 with a plug-in from Eksigent and the amaZon ion traps by trapControl 7.0, all from Bruker.

Data analysis

Organ and tissue proteomes were contrasted using compareMS2 exactly as previously described (Palmblad and Deelder, 2012) using MGF files generated by DataAnalysis (Bruker), each containing 2,000 tandem mass spectra. The compareMS2 output, or fraction of shared tandem mass spectra between each pair of datasets, was combined into a distance matrix in the MEGA format and a UPGMA tree generated by MEGA 5.05 (Tamura *et al.*, 2011) as presented in Figure 1.

Results and discussion

We describe the application of a recently developed method for molecular phylogenetics to species recognition and product authentication. The method selects the best tandem mass spectra (here 2,000/sample) and based on a pairwise comparison of individual spectra and LC-MS/MS datasets, a phylogenetic tree was generated using compareMS2 and MEGA 5.05 – both freely available software.

From Figure 1a it is clear that the described workflow successfully distinguishes the 17 different fish species in groups of closely related families, based on a small piece of muscle tissue. The 17 different species represented in the study are not all fully sequenced, illustrating a major advantage of the method in that it is not depended on any sequence database search. In fact, no peptide identification was used in the analysis. This makes the method equally applicable to

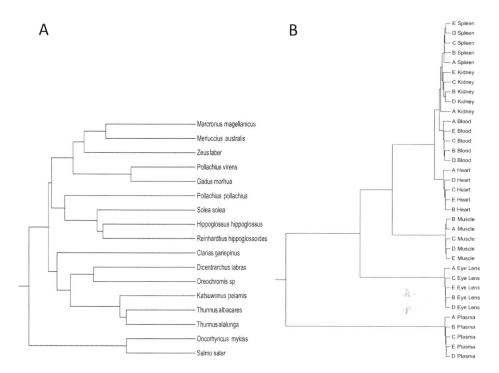

Figure 1. Species recogntion and phenotyping by tandem mass spectrometry. Proteomes were contrasted using compareMS2 based on MGF files generated by DataAnalysis (Bruker), each containing 2,000 tandem mass spectra. The compareMS2 output, or fraction of shared tandem mass spectra between each pair of datasets, was combined into a distance matrix in the MEGA format and a UPGMA tree generated by MEGA 5.05. (A) Molecular phylogeny of 17 different fish species. (B) Phenotyping of seven different tissue types from 5 rainbow trouts.

sequenced as well as non-sequenced organisms. Another potential advantage of the presented method is the unbiased use of all acquired data. The 2,000 automatically grouped and selected tandem mass spectra represents the bulk of high-quality spectra, compared to other published methods that uses manually selected spectra or peptides as molecular makers for species identification. By comparing different samples based on thousands of tandem spectra the method is essentially making comparisons based on the overall proteome of the different samples. This means that the method is also well suited for separating samples based on their proteomic phenotype, i.e. what proteins are expressed. In Figure 1b seven different organs and tissues from five adult rainbow trout are successfully clustered by compareMS2 and MEGA based on this proteomic phenotype. This is also an indication of the robustness of the method since all the 35 different samples (seven tissues from each of five fish) were correctly assigned to their organ/tissue group. The robustness of the method is also demonstrated by the fact that all data was processed automatically in exactly the same manner as in the previously published primate phylogenetic study, requiring no manually interpretation, selection of spectra or fine-tuning of parameters at any step of the workflow.

Acknowledgment

This work was supported by the Danish Strategic Research Council grant 3304-FVFP-08-K-08 and COST Action FA1002 and the Dutch Organization for Scientific Research (NWO) Vidi grant VI-917.11.398.

References

Berrini, A., Tepedino, V., Borromeo, V. and Secchi, C., 2006. Identification of freshwater fish commercially labelled 'perch' by isoelectric focusing and two-dimensional electrophoresis. Food Chemistry 96: 163-168.

Carrera, M., Canas, B., Pineiro, C., Vazquez, J. and Gallardo, J.M., 2007. De novo mass spectrometry sequencing and characterization of species-specific peptides from nucleoside diphosphate kinase B for the classification of commercial fish species belonging to the family Merlucciidae. Journal of Proteome Research 6: 3070-3080.

Mazzeo, M.F., De Giulio, B., Guerriero, G., Ciarcia, G., Malorni, A., Russo, G.L. and Siciliano, R.A., 2008. Fish Authentication by MALDI-TOF Mass Spectrometry. Journal of Agricultural and Food Chemistry 56: 11071-11076.

Palmblad, M. and Deelder, A.M., 2012. Molecular phylogenetics by direct comparison of tandem mass spectra. Rapid communications in mass spectrometry: RCM 26: 728-732.

Tamura, K., Peterson, D., Peterson, N., Stecher, G., Nei, M. and Kumar, S., 2011. MEGA5: molecular evolutionary genetics analysis using maximum likelihood, evolutionary distance, and maximum parsimony methods. Molecular biology and evolution 28: 2731-2739.

MALDI imaging mass spectrometry of ageing and osteoarthritic cartilage

Mandy J. Peffers[1], Berta Cillero-Pastor[2] and Gerd Eijkel[2]
[1]*Institute of Ageing and Chronic Disease, University of Liverpool, Liverpool, United Kingdom;*
peffs@liv.ac.uk
[2]*AMOLF-FOM Institute, Amsterdam, the Netherlands*

Aim

In this study we aimed to establish the spatial distribution of peptides in cartilage ageing and osteoarthritis in order to determine changing molecular events distinct between ageing and disease.

Introduction

Osteoarthritis (OA) is an age related joint disease characterized by a loss of cartilage extracellular matrix (ECM) (Goldring, 2000). Progressive destruction of articular cartilage is a hallmark of OA, leading to chronic pain and lameness. Indeed age is the most common risk factor for its initiation and progression with symptomatic OA affecting 10-20% of people aged over 50 years (Lawrence *et al.*, 2008). The explanation for this is an accumulation of 'wear and tear' injuries due to mechanical loading over the years. Although much work has been undertaken to investigate the pathogenesis of OA the molecular mechanisms involved are not fully understood, with few validated markers for disease diagnosis and progression being available.

Mass spectrometry (MS) is an analytical tool that enables the accurate measurement of both mass and charge of molecules. Matrix assisted laser desorption/ionization imaging mass spectrometry (MALDI IMS) of tissue samples is a powerful technique that allows for spatially resolved, comprehensive and specific characterization of hundreds of unknown molecular species (proteins, peptides, lipids and metabolites) *in-situ* in a single molecular imaging experiment (Seeley and Caprioli, 2008).

Methodologies which permit the study in detail of aging and OA cartilage organisation such as MALDI IMS, could improve the knowledge of how aging increases the risk of OA.

Methods

Full thickness equine cartilage slices from young (4 years old), old (greater than 15 years old) and OA old donors were removed from the mid condyle region of metacarpophalangeal joints collected from an abattoir. Samples were cut in triplicate obtaining 12μm thick sections and deposited on glass slides. Macroscopic and histological evaluation (following haematoxylin-eosin staining) of all samples was undertaken using Kawcak (Kawcak *et al.*, 2008) and modified Mankin (McIlwraith *et al.*, 2010) grading respectively. Following sample washing trypsin was deposited using a chemical inkjet printer. The samples were incubated overnight at 37 °C and a vibrational sprayer used to apply the matrix solution; α-Cyano-4-hydroxycinnamic acid (10 mg/ml) in 50% acetonitrile, 50% trifluoroacetic 0.1% (1:1 v/v).

A MALDI SYNAPT™HDMS system (Waters Corporation) instrument was used to perform the imaging experiments in the positive V-reflectron mode. The data analysis workflow included a number of AMOLF in-house software programs to undertake Principal Component Analysis (PCA) and Discriminant Analysis (DA) in order to interpret data generated. The first 20 principle components were used to probe spectral similarities and differences between young, old, and OA samples. Biomap 3.7.5.5 software (Novartis Pharma AG, Basel, Sweden) was used to generate ion images and quantify the intensity of different peptides. Peptide identification was undertaken using the MASCOT algorithm searching against the Swissprot and Unihorse databases following profiling MS/MS and imaging MS/MS experiments.

Results

DA was performed on the first 20 principle components in the dataset in order to remove noise and provide a conservative approach to data analysis. Three donors for each condition were examined in duplicate (young and old) or triplicate (normal and OA) technical replicates per group. PCA analysis, which is unsupervised, revealed differences between each sample type and resulting discriminant functions (DFs) classified the data in three groups: young, old and OA (Figure 1). Thus performing DA we demonstrated that the profiles of young, old and OA

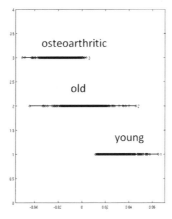

Figure 1. PCA analysis revealed differences between each sample type and resulting discriminant functions classified the data into young, old and osteoarthritic. Data is representative of DF 1. The positive part of the graph corresponds to young cartilage and the negative to osteoarthritic.

cartilages can be differentiated using the peptide profile. The scaled spectra represented better the raw intensity so was used to identify peaks for the MS/MS fragmentation and identification.

The presence of the ECM protein fibromodulin, collagen type II, matrilins-3 and cartilage oligomeric matrix protein were amongst those characterised in young, old and OA cartilage by MALDI-IMS. We observed a difference in the normalized intensity of numerous peptide peaks. The differences in the distribution of a number of peptides including a peptide with m/z 1,349.6 equivalent to a fibronectin peptide previously identified as a marker of OA in man (Cillero-Pastor et al., 2012) and a fibromodulin peptide ELHLDHNQISR with m/z 1,361.7, were identified between young, old and OA cartilages using Biomap software. By plotting the number of pixels against the average intensity observed differences for numerous peptides were identified. For example the highest intensity differences for the peptide at m/z 1349.6 were between young and OA samples and for the peptide m/z 1,361.7 were between old and OA cartilage. Additionally novel tentative OA markers were observed including the peptide m/z 1366.5 which was found to be significantly imaged at greater intensities in all OA samples (mean intensity 1.00±0.03, $P<0.001$).

Discussion

In this study we utilised recently implemented novel techniques to identify and spatially resolve peptides in ageing and OA cartilage. In addition to a number of potential age-related and disease specific peptide markers identified, a peptide with an m/z 1,361.7 previously identified as the fibromodulin peptide ELHLDHNQISR, was significantly reduced in OA samples. Interestingly our previous studies have identified a potential ADAMTS4 cleavage site in this peptide at $Asn^{167}–Gln^{168}$ (unpublished data). Although fibromodulin has been previously identified as a substrate for ADAMTS4 (Gendron et al., 2007), this was at the $Tyr^{44}–Ala^{45}$ bond (Fushimi et al., 2008).

Degraded fragments of the core fibromodulin protein have been observed in OA cartilage (Roughley et al., 1996) and with age (Cs-Szabo et al., 1995). This illustrates the potential usefulness of MALDI-IMS in identifying and spatially resolving novel cleavage sites with pathological relevance. Indeed this study would indicate that cleavage of the peptide ELHLDHNQISR is disease and not age related. Insights such as this may aid in the understanding of the age-related but not age-distinct disease OA.

References

Cillero-Pastor, B., Eijkel, G.B., Kiss, A., Blanco, F.J. and Heeren, R.M., 2012. Matrix assisted laser desorption ionization imaging mass spectrometry: A new methodology to study human osteoarthritic cartilage. Arthritis Rheum.

Cs-Szabo, G., Roughley, P.J., Plaas, A.H. and Glant, T.T., 1995. Large and small proteoglycans of osteoarthritic and rheumatoid articular cartilage. Arthritis Rheum 38: 660-668.

Fushimi, K., Troeberg, L., Nakamura, H., Lim, N.H. and Nagase, H., 2008. Functional differences of the catalytic and non-catalytic domains in human ADAMTS-4 and ADAMTS-5 in aggrecanolytic activity. J Biol Chem 283: 6706-6716.

Gendron, C., Kashiwagi, M., Lim, N.H., Enghild, J.J., Thogersen, I.B., Hughes, C., Caterson, B. and Nagase, H., 2007. Proteolytic activities of human ADAMTS-5: comparative studies with ADAMTS-4. J Biol Chem 282: 18294-18306.

Goldring, M.B., 2000. The role of the chondrocyte in osteoarthritis. Arthritis Rheum 43: 1916-1926.

Kawcak, C.E., Frisbie, D.D., Werpy, N.M., Park, R.D. and McIlwraith, C.W., 2008. Effects of exercise vs experimental osteoarthritis on imaging outcomes. Osteoarthritis Cartilage 16: 1519-1525.

Lawrence, R.C., Felson, D.T., Helmick, C.G., Arnold, L.M., Choi, H., Deyo, R.A., Gabriel, S., Hirsch, R., Hochberg, M.C., Hunder, G.G., Jordan, J.M., Katz, J.N., Kremers, H.M. and Wolfe, F., 2008. Estimates of the prevalence of arthritis and other rheumatic conditions in the United States. Part II. Arthritis Rheum 58: 26-35.

McIlwraith, C.W., Frisbie, D.D., Kawcak, C.E., Fuller, C.J., Hurtig, M. and Cruz, A., 2010. The OARSI histopathology initiative – recommendations for histological assessments of osteoarthritis in the horse. Osteoarthritis Cartilage 18 Suppl 3: S93-105.

Roughley, P.J., White, R.J., Cs-Szabo, G. and Mort, J.S., 1996. Changes with age in the structure of fibromodulin in human articular cartilage. Osteoarthritis Cartilage 4: 153-161.

Seeley, E.H. and Caprioli, R.M., 2008. Molecular imaging of proteins in tissues by mass spectrometry. Proc Natl Acad Sci U S A 105: 18126-18131.

Protein expression in bovine mononuclear cells after stimulation with lipopolysaccharides and lipoteichoic acid: a proteomic approach

Laura Restelli[1], Tommaso Serchi[2], Nicola Rota[3], Cristina Lecchi[3], Jenny Renaut[2] and Fabrizio Ceciliani[3]

[1]*Department of Veterinary Science for Animal Production, Food Safety and Animal Health; Faculty of Veterinary Medicine – Università degli Studi di Milano, Milan, Italy; laura.restelli@unimi.it*

[2]*Department of Environment and Agro-Biotechnologies; Centre de Recherche Public – Gabriel Lippmann, Belvaux, Luxembourg*

[3]*Department of Veterinary Science and Public Health; Faculty of Veterinary Medicine – Università degli Studi di Milano, Milan, Italy*

Objectives

The aim of this project was to investigate the proteome of bovine mononuclear cells in physiological condition and after stimulation with different pathogen associated molecular patterns.

Escherichia coli and *Staphylococcus aureus* infections induce different protein expression patterns during intra-mammary infections. Mononuclear cell population include both lymphocyte and monocyte cell types. The study of mononuclear protein expression during inflammation can be of great help in understanding the mechanism underlying innate immune response against pathogens.

On the background of the profound relationship between monocytes and lymphocytes, which develop an intense cross-talking after inflammatory activation, the aim of the present work is to provide new information regarding mononuclear cells proteome profile during inflammatory processes after *in vitro* experimental challenge with Gram+ and Gram- derived PAMPs (lipoteichoic acid and lipopolysaccharides respectively). These molecules are known to activate monocytes, and our strategic aim is to investigate how the whole mononuclear population proteome may change after the activation of just one population.

Materials and methods

Blood sampling, cells isolation, cells stimulation and cells lysis

Blood was collected from four clinically healthy cows. The experiment was carried out after incubation of the whole population of mononuclear cells, i.e. monocytes and lymphocytes. After purification, cells were then co-coltured in 75 cm^2 flasks in RPMI-1640 medium with additions of 20 mM Hepes, 10% Fetal Bovine Serum (FBS), 100 IU/ml penicillin, 100 μg/ml streptomycin, 1% non-essential aminoacids (Sigma Aldrich, Italy). After two hours of resting at 37 °C in a humidified atmosphere of 5% of CO_2 cells were stimulated with either 100 ng/ml LPS (Dilda *et al.*, 2012) or 5 μg/ml LTA (Meijerink *et al.*, 2011) and incubated for 24 h at 37 °C with 5% of CO_2. Cells incubated with RPMI 1640 without challengers' addition were used as negative controls.

After 24 hours of stimulation, proteins were extracted by adding lyses buffer containing 7 M Urea, 2 M Thiourea, 2% CHAPS and 40 mM Tris (pH 8.5) (Sigma Aldrich, Italy), followed by 5 cylces of 30" sonication and 30" ice and centrifugation at 14,000 rpm (10 min). Protein concentration was determined by the Bradford method (595 nm) using 1 mg/ml of bovine serum albumin as standard.

2D-Different Gel Electrophoresis (DIGE), protein identification and analysis

The proteomic analysis was carried out at the Centre de Recherche Public – Gabriel Lippmann in Luxembourg in collaboration with Doctor Jenny Renaut by means of a COST-Short Term Scientific Mission Grant. Samples were analysed using a 2D-DIGE, following the procedure described by Kaulmann and co-workers (Kaulmann *et al.*, 2012).

Proteins for each sample (50 μg) were labeled with either Cy3 or Cy5 dyes (400 pmol). In addition, a pool containing equal amounts of each sample was labeled with Cy2 dye in order to be used as internal standard. The first-dimension isoelectric focusing was performed on 24 cm strips (GE Healthcare, France), with a 3-10 linear pH gradient. The second-dimension separation was performed on a 12.5% pre-cast polyacrylamide gel (Serva™), in an Ettan Dalt II system (GE Healthcare, France). Each gel was scanned at three different wavelengths corresponding to the different dyes and all the images were analysed using the Decyder 2D software 7.0 (GE healthcare, France). A 1-way ANOVA was performed to analyse the results and highlights protein of interest. A *P*-value below 0.05 (two-sided) was chosen to indicate significance. Furthermore, a hierarchical clustering analysis was performed using the Ward's linkage method.

The study is still ongoing. Protein identification will be achieved by MALDI-TOF/TOF mass spectrometry and bioinformatics tools such as the MASCOT software.

Results and discussion

From a practical point of view this experiment gave us the possibility to learn new useful information about the use of mononuclear cells in proteomics.

Monocytes didn't provide an adequate amount of protein for a 2D-DIGE analysis, even after doubling the amount of starting blood when cultured alone; after the co-incubation with lymphocytes and the reduction of the resting period from 24 hour to 2 hours, monocytes looked more vital.

The ionic strength in the samples was high, probably due to the utilization of PBS throughout washing of cultivated cells; the first dimension was therefore run at 30 V for almost 24 hours in order to eliminate the high amount of salts and keep the value I/strip (resistance per strip) under 50 μA, avoiding strips damages.

All the 2D gels show a vertical gap corresponding to the pH values 8-9. Three of the six gels were analysed, but excluded from the statistical analysis as they show a wider gap that results in the absence of several proteins. Gels will be performed again at a later stage in order to have a more robust statistical analysis.

After the 1-way ANOVA on the three remaining gels (corresponding to two samples per treatment/control), a picking list of 9 spots was generated, corresponding to proteins differentially regulated in their expression after samples treatment with lipoteichoic acid (LTA) or lipopolysaccharides (LPS) (Figure 1a-b). Table 1 shows that most of the spots follow the same trend after stimulation with the challengers: indeed, 5 of the 9 spots are down-regulated after both treatments if comparing to the control samples; whereas 3 of the 9 spots are up-regulated after the treatments. Only one spot seems to be poorly up-regulated after treatment with LPS, but down-regulated after treatment with LPS. The table also provides the LPS/LTA ratio.

Furthermore a dendrogram grouping the three different treatments according to their proteomic profile (considering the 9 selected spots) was generated by hierarchical clustering (see Figure 1c). The dendrogram showed a clear grouping in three clusters allowing the distinction of the two treatments according to the statistically significant protein expression changes. Likewise control and treated samples can be distinguished by their protein expression patterns.

Finally, the Principal Component Analysis confirmed the previous results, showing a clear difference among control, LPS-treated and LTA-treated samples (Figure 1d).

Lipoteichoic acid and lipopolysaccharides induce protein expression changes in mononuclear cells. Proteins generally follow the same trend after LPS or LTA stimulation and most of the selected spots show to be down-regulated after the treatments. Results are still preliminary, but will provide new interesting information about mononuclear cells proteome.

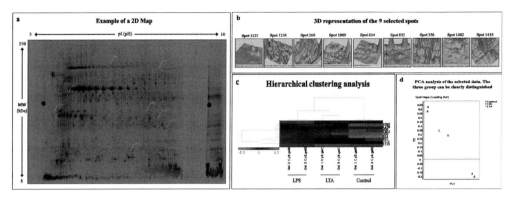

Figure 1. 2D Gel example map, selected spots and statistical analysis results. After the 1-way ANOVA, 9 spots were selected corresponding to proteins differentially regulated in their expression after samples treatment. Furthermore a hierarchical clustering analysis and a Principal Component Analysis (PCA) were performed on the selected proteins: (a) example of a 2D gel map, the 9 selected spots are highlighted; (b) 3D representation of the selected spots; (c) Hierarchical clustering analysis showing a clear grouping in three different clusters (LPS, LTA, control); (d) the PCA analysis shows a clear difference among control, LPS-treated and LTA-treated samples.

Table 1. List of the 9 selected spots and their expression trend in treated samples compared to controls. The P-value of the 1-way ANOVA is also provided.

Spot reference number	Ratio LPS/control	Ratio LTA/control	Ratio LPS/LTA	1-way ANOVA (P<0.05)
1238	(–) -2.28	(–) -1.65	(–) -1.38	0.0075
268	(–) -1.40	(–) -1.27	(–) -1.10	0.0079
1089	(–) -1.43	(–) -1.13	(–) -1.27	0.011
350	(–) -1.50	(–) -1.33	(–) -1.13	0.037
1802	(–) -1.46	(–) -1.31	(–) -1.12	0.041
814	(+) 2.49	(+) 2.05	(+) 1.21	0.032
832	(+) 1.60	(+) 1.26	(+) 1.27	0.034
1416	(+) 1.29	(+) 1.02	(+) 1.25	0.045
1123	(+) 1.05	(–) -1.09	(+) 1.15	0.00018

References

Dilda, F., Gioia, G., Pisani, L., Restelli, L., Lecchi, C., Albonico, F., Bronzo, V., Mortarino, M. and Ceciliani, F., 2012. Escherichia coli lipopolysaccharides and Staphylococcus aureus enterotoxin B differentially modulate inflammatory microRNAs in bovine monocytes. Vet J 192: 514-516.

Kaulmann, A., Serchi, T., Renaut, J., Hoffmann, L. and Bohn, T., 2012. Carotenoid exposure of Caco-2 intestinal epithelial cells did not affect selected inflammatory markers but altered their proteomic response. Br J Nutr 108: 963-973.

Meijerink, M., Ulluwishewa, D., Anderson, R.C. and Wells, J.M., 2011. Cryopreservation of monocytes or differentiated immature DCs leads to an altered cytokine response to TLR agonists and microbial stimulation. J Immunol Methods 373: 136-142.

The 'hidden' proteome of cow's and Jennys' milk as revealed by combinatorial peptide ligand libraries

Elisa Fasoli[1], Alfonsina D'Amato[1], Vincenzo Cunsolo[2], Attilio Citterio[1] and Pier Giorgio Righetti[1]
[1]Laboratory of Proteomics, Department of Chemistry, Material and Chemical Engineering, Politecnico di Milano, Milano, Italy; elisa.fasoli@polimi.it
[2]Department of Chemical Sciences, University of Catania, Viale A. Doria 6, 95125 Catania, Italy

Objectives

Milk is the most important food for young mammals and a common source of proteins and microelements for adults. In addition, it is a relevant means for transferral of immunity to pathogens from the mother to the newborn, as it contains antimicrobial and immunomodulatory proteins that are active in the digestive tract of newborns. Different kinds of milk are present in human alimentation: the most popular being cow's milk, but also Jennys and goats are of importance, especially in presence of milk allergy.

Cow's milk is of great human nutritional and economic significance, yet its minor protein repertoire has not been characterized in depth. Mature bovine milk contains about 3.3% protein, of which about 80% consists of caseins (CN), namely α_{S1}-, α_{S2}-, β- and κ-caseins, the remaining 20% being serum albumin, β-lactoglobulin, α-lactalbumin and other low-abundance proteins (Swaisgood, 1995).

Jennys' milk is today categorized among the best mother milk substitute for allergic newborns, due to its much reduced or absent allergenicity, coupled to excellent palatability and nutritional value. However, to present, only a handful of proteins has been characterized, just about the standard eight to ten major ones known in all types of milk.

By exploiting the combinatorial peptide ligand library technology, and treating large volumes (up to 300 ml) of defat, de-caseinized (whey) milk, we have been able to identify a large number of unique gene products, by far the largest description so far of this precious nutrient.

In fact, the aim of this study was firstly to characterize the minor components in these complex protein systems, that means cow's and donkey's milk, by the application of the combinatorial peptide ligand libraries technology (Boschetti and Righetti, 2008; Righetti *et al.2010*) and by mass spectrometry analysis, and secondly to investigate the presence of any possible allergens in cow's milk.

Material and methods

Milk samples were centrifuged (2,000×g, 20 min) at 4 °C and the fat layer (with the milk fat globules) carefully removed. The fat-free solution was then ultracentrifuged (100,000×g, 1 h) so as to precipitate the casein micelles. The clear supernatant (whey fraction) was mixed with the Roche complete protease inhibitor and incubated with combinatorial peptide ligand libraries, commercially available as ProteoMiner (Bio-Rad). The captured proteins were desorbed using different eluent solutions (Candiano *et al.*, 2009) and the eluted proteins loaded in double onto SDS-PAGE gels for separation by mono-dimensional electrophoresis. Each sample was mixed with 10 μl of Laemmli buffer (4% SDS, 20% glycerol, 10% 2-mercaptoethanol, 0.004% bromophenol blue and 0.125 M Tris HCl, pH approx. 6.8) and heated in boiling water for 5 min and immediately loaded in the gel. The SDS-PAGE slab was composed by a stacking gel (125 mM Tris-HCl, pH 6.8, 0.1% SDS) with a large pore polyacrylamide (4%) cast over the resolving gel (8-18% acrylamide gradient in 375 mM Tris-HCl, pH 8.8, 0.1% SDS buffer). The cathodic and anodic compartments were filled with Tris-glycine buffer, pH 8.3, containing 0.1% SDS. Electrophoresis was at 100 V until the dye front reached the bottom of the gel. Staining and distaining of one half of gels were performed with Colloidal Coomassie Blue and 7% acetic acid in water, respectively. The SDS-PAGE gels were scanned with a VersaDoc imaging system (Bio-Rad). The gels were cut, digested by trypsin to allow protein identification by nanoLC MS/MS with an LTQ-Orbitrap mass spectrometer (ThermoScientific, Bremen, Germany) equipped with a nanoelectrospray ion source (Proxeon Biosystems, Odense, Denmark) for cow's milk and by RP-HPLC/nESI-MS/MS analysis with a linear ion trap nano-electrospray mass spectrometer (LTQ, Thermo Fischer Scientific, San Jose, CA) in the case of Jennys' milk. For both spectrometers the various sample lanes of SDS-PAGE gels were cut in slices of about 0.5 cm along the migration path, and proteins were reduced by 10 mM DTT and alkylated by 55 mM iodoacetamide. The gel pieces were shrunk in acetonitrile and dried under vacuum; proteins were digested overnight with bovine trypsin. The tryptic mixtures were acidified with formic acid up to a final concentration of 1% and injected in a capillary chromatographic system.

The other half of gels was blotted on to PVDF or nitrocellulose membranes, which were incubated with sera of milk allergic patients in order to detect all possible milk allergens.

The desired volume of each non-treated sample and CPLL eluates was explored in a two-dimensional (2D) mapping study after solubilisation in 2-D sample buffer (7 M urea, 2 M thiourea, 3% CHAPS, 40 mM Tris). After reduction of disulphide bridge reduction by adding 5mM TCEP [Tris(2-carboxyethyl)phosphine hydrochloride] and the alkylation of reduced –SH groups by adding 150 mM De-Streak [Bis-(2-hydroxyethyl)disulphide, $(HOCH_2CH_2)_2S_2)$], all samples were adsorbed on to 7 cm long IPG strips (Bio-Rad), pH 3-10 l, for the isoelectric focusing (IEF). IEF was carried out with a Protean IEF Cell (Bio-Rad Laboratories) in a linear voltage gradient from 100 to 1000 V for 5 h, 1000 V for 4 h, followed by an exponential gradient up to 5,000 V, for a total of 25 kV/h. For the second dimension, the IPGs strips were equilibrated for 25 min in a solution containing 6 M urea, 2% SDS, 20% glycerol, 375 mM Tris-HCl (pH 8.8) under gentle shaking. The IPG strips were then laid on 8-18% acrylamide gradient SDS-PAGE gel slab with 0.5% agarose in the cathodic buffer (192 mM glycine, 0.1% SDS and Tris-HCl to pH 8.3). The electrophoretic run was at 5 mA/gel for 1 h, followed by 10 mA/gel for 1 h and 15 mA/gel until the dye front reached the gel bottom. The 2-DE gels, stained by Colloidal Coomassie Blue solution, were scanned with a Versa-Doc image system (Bio-Rad), by fixing the acquisition time at 10 sec; the relative gel images were captured via the PDQuest software (Bio-Rad Laboratories).

Results and discussion

The use of combinatorial peptide ligand libraries allowed discovering and identifying a large number of previously unreported proteins in cow's and donkey's whey.

In particular as concern cow's milk, whereas comprehensive whey protein lists progressively increased in the last six years from 17 unique gene products to more than 100, our findings have considerably expanded this list to a total of 149 unique protein species, of which 106 were not described in previous proteomics studies (Figure 1A). As an additional interesting result, a polymorphic alkaline protein was observed with a strong positive signal when blotted from an isoelectric focusing separation in gel and tested with sera of allergic patients (Figure 1B). This polymorphic protein, found only after treatment with the peptide library, was identified as an immunoglobulin (IgG), a minor allergen that had been largely amplified.

Concerning Jennys' milk, we have been able to identify 123 unique gene products, most of which (82%) were exclusively found in the whey proteome after ProteoMiner treatment (Figure 1C). Among the 123 identified proteins, 23 components were the well known milk proteins (caseins, alpha-lactalbumin, beta-lactoglobulins, lysozyme, serum albumin and lactoferrin), whereas the others (83%) were not described in previous proteomic studies and therefore represent the most comprehensive list of donkey's whey proteins at present (Figure 1D). Due to poor knowledge of the donkey's genetic asset, only 11% of the proteins could be identified by consulting the data base of *Equus asinus*; the largest proportion (62%) could be identified by homology with the proteins of *Equus caballus*.

In conclusion, this first use of peptide library treatment of milk proteins allowed evidencing new species not yet known as milk proteins; additionally, among the proteins found, it was possible to identify minor allergens that generally are hidden by the immunological reactions generated by major allergenic antigens. Clearly, the amplification of low abundance species (as well engendered by combinatorial peptide libraries) is a way to detect allergens of very low concentration. Since the list of milk components reported in our study is by far the most comprehensive at present, it could serve as a starting point for the functional characterization of low-abundance proteins possibly having novel pharmaceutical, diagnostic and biomedical applications.

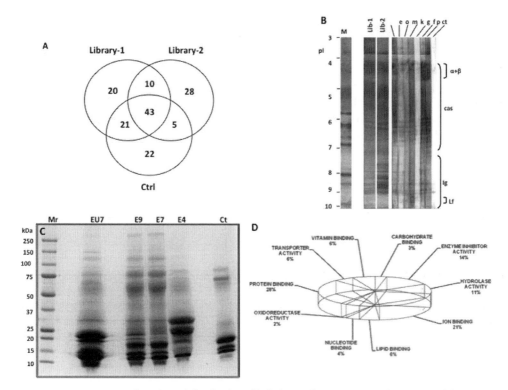

Figure 1. Proteomics of cow's and donkey's milk. (A) Overlapping Venn diagrams of the proteins detected in the initial milk whey material (Ctrl) and two eluates from Library-1 and Library-2. (B) Immunoblotting from an IEF analysis of milk whey library eluates against allergic patients' sera: lanes 'Lib-1' and 'Lib-2' are Coomassie stained eluates, the following lanes on the right are immunoblots of the same protein mixture with sera 'e', 'o', 'm', 'k', 'g', 'f' and 'p'. 'ct' is the negative control. Regions indicated by 'cas', 'α+β', 'Ig' and 'Lf' mean, respectively, where α-lactalbumin, β-lactoglobulin, immunoglobulins and lactoferrin are generally detected. (C) SDS-PAGE profiling of Jennys' milk. Order of loading from left to right: Mr: molecular mass markers; EU7: 4%SDS eluate of Jennys' milk treated by ProteoMiner beads at pH=7 (purified by ultracentrifugation); E9, E7, E4: 4%SDS eluates of Jennys' milk treated by ProteoMiner beads at pH 9.3, 7.2 and 4.0, respectively; Ctrl: control, untreated Jennys' milk. In all cases, a 25 µl volume was seeded into the gel pockets. Staining with Coomassie brilliant blue. (D) Gene Ontology (GO) functional classification of all 123 identified proteins in Jennys' milk.

Acknowledgements

Financial support was from Fondazione Cariplo (Milan), from PRIN 2009 (MURST, Rome) and by a grant from MURST (PRIN 2008, project number 20087ATS57).

References

Boschetti, E. and Righetti, P.G., 2008. The ProteoMiner in the proteomic arena: a non-depleting tool for discovering low-abundance species. J Proteomics 71: 255-264.

Candiano, G., Dimuccio, V., Bruschi, M., Santucci, L., Gusmano, R., Boschetti, E., Righetti, P.G. and Ghiggeri, G.M., 2009. Combinatorial peptide ligand libraries for urine proteome analysis: investigation of different elution systems. Electrophoresis 30: 2405-2411.

Righetti, P.G., Boschetti, E., Kravchuk, A.V. and Fasoli, E., The proteome buccaneers: how to unearth your treasure chest via combinatorial peptide ligand libraries. Expert Rev Proteomics 7: 373-385.

Swaisgood, H.E., 1995. In: Jensen, R.G. (ed.), Handbook of milk composition. Academic Press.

Comparison between Coomassie Blue and Silver staining in porcine saliva samples for proteomics: technical considerations: a preliminary experiment

María Fuentes-Rubio[1], Carlos de Torre[2], Ana Gutiérrez[1], Jose J. Cerón[1] and Fernando Tecles[1]
[1]*Department of Animal Medicine and Surgery, Regional Campus of International Excellence 'Campus Mare Nostrum', University of Murcia, 30100, Espinardo, Murcia, Spain; mfr2@um.es*
[2]*Proteomics Unit, Universitary Hospital Virgen de la Arrixaca, Ctra Madrid-Cartagena s/n 30120 El Palmar-Murcia, Spain*

Objectives

Whole saliva is a complex physicochemical fluid, composed of secretions from the different salivary glands. The variable and complex composition of saliva entails a number of difficulties when it is used as a biological sample under investigation by proteomics. In a proteomics study, after two-dimensional (2DE) electrophoresis, the selection of the appropriate protein detection method is highly important (Westermeier and Marouga, 2005). 'Detection' is often synonymous with staining (Miler *et al.*, 2006).

The principal features of the ideal detection method (or type of staining) should be reproducibility and repeatability, ability of being compatible with mass spectrometry (MS), having enough sensitivity for low copy number proteins and having a wide dynamic range (wide linear relationship between the quantity of protein and the staining intensity). Moreover the method should be easy and fast to perform, non-toxic, environment-friendly and affordable (Westermeier and Marouga, 2005).

The aim of the report was to review the requirements and procedures needed for treat and staining of proteomic saliva samples to show our personal experience with the two most used stains in proteomics (Coomassie Brilliant Blue and Silver Nitrate). To authors' knowledge, one of them has never been used in saliva (Coomassie Brilliant Blue).

Material and methods

The samples used in this report were no stimulated whole saliva samples, which were obtained by introducing a small sponge in the pigs' mouth for at least one minute with the help of a metal rod as previously described (Fuentes *et al.*, 2011). The sponges were placed in collection devices and were centrifuged at 4,000×*g* for eight min to obtain saliva. Saliva samples were kept at -80 °C until analysis. Healthy 3-4 months of age crossbred growing pigs ([*Sus scrofa domesticus*] Duroc × [Landrace × Large White]) from the experimental farm unit of the University of Murcia (Spain) were used in this study.

Salivary concentration was determined by BioRad DC protein assay (BioRad, Hercules, CA, USA), a modification of Lowry assay (Lowry *et al.*, 1951). In order to obtain salivary 2DE gels, commercial immobilized pH gradient (IPG) strips were used (GE Healthcare Europe GmbH, Freiburg, Germany) and the re-hydration step was made according to manufacturer's procedure. Proteins were focused in the first dimension in the Ettan IPGphor 3 system (GE Healthcare Europe GmbH, Freiburg, Germany). For the second dimension, reduced and alkylated strips were subjected to SDS-PAGE in homemade 12% polyacrylamide gels in a vertical electrophoresis chamber (Ettan DALT six Electrophoresis System, GE Healthcare Europe GmbH, Freiburg, Germany). Two of the most used techniques for total protein staining were used in this study: Coomassie Brilliant Blue (CBB) and Silver Nitrate (SN) staining.

Procedure A: CBB staining

Saliva samples were treated prior to electrophoresis with 2-D Clean-Up Kit (GE Healthcare Europe GmbH, Freiburg, Germany) and then, the protein concentration of each sample was quantified again using 2-D quant kit. Gels were fixed and stained with Coomassie brilliant blue R-250 according to staining protocol (Neuhoff *et al.*, 1988).

Procedure B: SN staining

An aliquot of 60 μg of protein from whole saliva sample was lyophilized to reduce sample volume. Gels were stained according to silver staining protocol (Miller *et al.*, 2009).

Results and discussion

Comparison of the results obtained with different dyes has several limitations. One that can be found is that silver staining often produces a pattern different from the one achieved with Coomassie blue (Görk *et al.*, 2000). However, the approach of this report allows us to have an idea about the advantages, disadvantages and the requirements of one or other staining protocol in saliva samples.

The main aspects in which differences have been found were as listed below.

Pre-treatment of saliva samples

The characteristics of the saliva samples made necessary a sample pre-treatment in both procedures. For the CBB stain, pre-treatment is more aggressive than in the case of SN stain. The high volume needed in the CBB staining made necessary a precipitation with 2-D Clean Up kit to avoid problems of high conductivity, high levels of interfering substances or low protein concentration. For this purpose, a commercial kit was used (Yang *et al.*, 2006). Similarly, samples were pooled (each sample contributed to make the pool with an equal amount of protein) to avoid the need to use a large volume of a single sample in order to reach the minimum concentration required for that type of staining. Thus, multiple saliva samples with smaller volume of each one were used instead of a large volume of a unique sample. This initial step of CBB staining might involve a loss of information. Either because it loses some protein by an irreversible precipitate, or because using pools in proteomic studies could result in some loss of information in comparison to the use of single samples (Zolg, 2006). On the other hand, samples were only lyophilized when SN staining was used (Gutiérrez *et al.*, 2013), by using less amount of protein per sample, and thus less volume, the presence of interfering substances decreased.

Stain procedure

The CBB staining is simpler to perform than the SN staining. While it is true that the CBB stain is slow and prolonged in time (especially the distained step), this procedure does not require much precision as for the duration of each step than in the case of SN stain. The silver nitrate staining steps have to be prolonged and there is a loss of sensitivity (Westermeier and Marouga, 2005). The major difficulties in both procedures appear in the distaining or developing step. Distaining of CBB stained gels has to be done carefully, because during distaining with the alcohol-containing solution the protein spots are partly distained as well. Some proteins lose the dye before the background of the gel is distained (Westermeier *et al.*, 2008). If the clearance is the problem of the Coomassie stained gels, darkening is the problem of gels stained with silver nitrate. The developing of silver nitrate stained gels is a very subjective process, which distained times are approximate. The person processing the gels should decide the optimal time point to stop the reaction, trying to achieve a balance between stain of spot and background displayed by darkening the gel. Because it is not an endpoint procedure, the staining intensities can vary from gel to gel and this results in decreased reproducibility (Miler *et al.*, 2006; Westermeier and Marouga, 2005). In our experience, different CBB stained gels had a more similar pattern among them that the gels stained with SN protocol, where more variability could be observed. Thus, at this point, silver nitrate staining has more disadvantages than CBB staining because, while an over-distained gel with Coomassie may be re-stained, there is no way back in the case of a gel too stained with silver nitrate. As is the case of other types of samples, at first glance, the spots seem to be more stained with SN protocol than with CBB stain; however, this further enhances the potential artefacts present in the gel. Moreover, the gel stained with CBB has less background than SN stained gel, which eases good exposure of spots with smaller amounts of protein.

MS compatibility

CBB stained gels do not represent any compatibility problem with MS. However, for doing MS analysis in silver stained gels, a modified protocol is needed. The SN staining protocol can be modified by omitting glutardialdehyde from the sensitizing solution and formaldehyde from the silver solution (Yan *et al.*, 2000), but the detection of sensitivity decreases with this step (Westermeier *et al.*, 2008).

In view of the data collected, saliva does not behave very differently from other samples in terms of gel staining for proteomic studies. Thus, if the CBB staining is chosen to ensure good reproducibility and compatibility with MS, stronger sample pre-treatment is necessary. In contrast, when SN staining is preferred to guarantee high sensitivity, the process after electrophoresis is more elaborate and requires modifications to the original protocol to safeguard compatibility with MS at the expense of reproducibility. In summary, the choice of either staining procedure depends on the importance we give to each one of the advantages held by each of the methods.

References

Fuentes, M., Tecles, F., Gutiérrez, A., Otal, J., Martínez-Subiela, S. and Cerón, J., 2011. Validation of an automated method for salivary alpha-amylase measurements in pigs and its application as a biomarker of stress. Journal of veterinary diagnostic investigation 23: 282-287.

Görk, A., Obermaier, C., Boguth, G., Harder, A., Scheibe, B., Wildgruber, R. and Weis, W., 2000. The current state of two-dimensional electrophoresis with immobilized pH gradients. Electrophoresis 21: 1037-1053.

Gutiérrez, A., Nöbauer, K., Soler, L., Razzazi-Fazeli, E., Gemeiner, M., Cerón, J. and Miller, I., 2013. Detection of potential markers for systemic disease in saliva of pigs by proteomics: A pilot study. Veterinary immunology and immunopathology 151: 73-82.

Lowry, O., Rosebrough, N., Farr, L. and Randall, R., 1951. Protein measurement with the folin phenol reagent. The journal of biological chemistry 193: 265-275.

Miler, I., Crawford, J. and Gianazza, E., 2006. Protein stains for proteomic applications: Which, when, why? Proteomics 6: 5385-5408.

Miller, I., Wait, R., Sipos, W. and Gemeiner, M., 2009. A proteomic reference map for pig serum proteins as a prerequisite for diagnostic applications. Research in Veterinary Science 86: 362-367.

Neuhoff, V., Arold, N., Taube, D. and Ehrhardt, W., 1988. Coomassie Blue Colloidal staining. Electrophoresis 9: 255-262.

Westermeier, R. and Marouga, R., 2005. Protein detection methods in proteomics research. Bioscience Reports 25: 19-32.

Westermeier, R., Naven, T. and Höpker, H.-R., 2008. Proteomics in Practice. A Guide to Successful Experimental Design. Wiley-VCH Verlagsgesellschaft GmbH, Weinheim, Germany, 482 pp.

Yan, J., Wait, R., Berkelman, T., Harry, R., Westbrook, J., Wheeler, C. and Dunn, M., 2000. A modified silver staining protocol for visualization of proteins compatible with matrix-assisted laser desorption/ionization and electrospray ionization-mass spectrometry. Electrophoresis 21: 3666-3672.

Yang, L., Liu, X., Cheng, B. and Li, M., 2006. Comparative analysis of whole saliva proteomes for the screening of biomarkers for oral lichen planus. Inflammation Research 55: 405-407.

Zolg, W., 2006. The proteomic search for diagnostic biomarkers. Lost in traslation? Molecular & Cellular Proteomics 5: 1720-1726.

Isoelectric focusing in characterisation of alkaline phosphatase isozyme from bovine nasal secretion and mucosa

M. Faizal Ghazali[1], Nicholas N. Jonsson[2], M. McLaughlin[2], I. Macmillan[2] and P. David Eckersall[1]
[1]Institute of Infection, Inflammation and Immunity, University of Glasgow, United Kingdom; m.ghazali.1@research.gla.ac.uk
[2]School of Veterinary Medicine, University of Glasgow, United Kingdom

Introduction

Investigation of bovine nasal secretion has revealed a higher activity of alkaline phosphatase (ALP) than that found in serum. However it is not clear whether ALP is synthesized in the nasal tissue and the type of ALP isozyme that is produced in the secretion. The aim of this study was to measure the level of ALP activity and to separate the ALP isozymes in bovine nasal secretion and in extracts of relevant bovine tissues by isoelectric focusing (IEF).

Material and methods

Nasal secretions (NS) were collected from 38 clinically healthy Holstein-Friesian cows aged 2-5 years on the University of Glasgow Cochno Farm as previously described (Ghazali *et al.*, 2012). Tissue samples were obtained *post mortem* from 6 further cows (Holstein-Friesian, 2-6 years) free from infectious and respiratory disease. Two grams of nasal mucosa, small intestine, heart, liver and kidney, obtained from each animal were washed with isotonic saline solution and homogenized mechanically in 5 ml saline containing 20% (v/v) n-butanol. The ALP activity in tissue extractions were measured using para-nitrophenyl phosphate (pNPP) enzymatic reaction. Histochemistry of the bovine nasal mucosa was performed using snap freezing technique, tissue was dissected at 0.7 μm, were fixed with acetone and stained with Vector® Blue Alkaline Phosphatase Substrate Kit (Vector Labs, California, USA).

Isoelectric focusing for the separation of ALP isozymes were undertaken with Invitrogen Novex® pH 3-7 IEF Gel (Life Technologies, Carlsbad, California, USA). Bovine nasal secretion and aliquots of tissue extraction from nasal mucosa, small intestine, heart, liver and kidney were selected randomly for the separation of ALP isozymes. Samples were prepared according to manufacturer's instructions. 10 μl of the prepared samples were loaded into each well. IEF was conducted at 20 °C using the following voltage gradient: 100 V, 1 hour; 200 V, 1 hour; 500 V, 30 minutes. After focusing, the gel was stained with Pierce 1-Step™ NBT-BCIP ALP substrate solutions (Thermo Fisher Scientific Inc, Illinois, USA) incubated at 37 °C up to 5 hours. Isoelectric points (pI) of the isozymes were determined using IEF protein markers of know pIs (IEF Markers 3-10, SERVA liquid mix (Invitrogen AB, Lidingö, Sweden).

Results

Five to 12 ml volumes of nasal secretion were collected per tampon from the cows. Protein concentrations in bovine nasal secretion ranged from 8.21 to 33.7 g/l. ALP concentrations in the bovine nasal secretion were up to 10 fold higher than the reference range for serum from healthy cattle (Table 1). The nasal mucosa had the highest ALP activity in the tissue extracts suggesting that the ALP in the NS is locally produced. Histochemistry showed strong ALP activity in the nasal epithelium and serous glands.

The estimated pI values of ALP from nasal secretion and tissue extracts are given in Table 2. Isoelectric focusing demonstrated that ALP bands from the nasal secretion and nasal mucosa are identical and resolved as a major band (pI 5.2) along with a complex of minor bands (pI 5.6-6.2) suggesting the presence of various minor ALP forms. The bovine bone ALP resolved

Table 1. Mean ALP activity extracts in bovine tissues (n=6) recorded as total units of enzyme activity per gram of tissue extracted.

Tissue	Units	Mean	Standard deviation (s)	Median	Range	Reference (bovine serum)
Nasal secretion	U/l	1192	500	1236	144-2,392	20-280[a]
Nasal mucosa	U/g	10.0	9.6	8.4	1.0-27.6	
Intestinal mucosa	U/g	8.3	8.7	2.3	2.2-24.0	
Heart	U/g	0.6	0.4	0.6	0.2-1.1	
Kidney	U/g	7.3	4.5	7.4	3.9-16.2	
Liver	U/g	9.5	10.0	7.0	1.04-27.2	

[a] Laboratory reference range.

Table 2. pI value estimates for ALP isozymes of nasal secretion and various tissues using Invitrogen Novex® pH 3-7 gradient IEF Gel.

ALP	pI value (Major bands)
Nasal secretion	5.2
Nasal mucosa	5.2
Liver	4.2
Bone	4.8
	6.5
Intestinal mucosa	6.0
Kidney	5.3

as a broad complex of bands (pI 4.8-6.5) consisting of two major bands at pI 4.8 and pI 6.5. The liver ALP migrated to the more anodal region (pI 4.2) suggesting a greater negative charge density than the other ALP extracts. Other ALP such as the Intestinal ALP were in the cathodic region, having a major band (pI 6.0) along with a complex of minor bands (pI 5.3-6.2) whereas kidney ALP resolved as a major band (pI 5.3) along with several minor bands (pI 5.3-6.0). The differences of ALP pI of the nasal secretion and nasal mucosa from other tissue extracts, suggesting that they are of different isozymes (Figure 1).

Conclusion

High concentration of ALP in the nasal secretion and nasal mucosa extraction along with the histochemistry findings have confirmed that the nasal ALP results from local synthesis and secretion of the enzyme. Isoelectric focusing of ALP from the bovine tissues indicates that the ALP from the nasal mucosa has a different isozyme from the other tissues such as liver or intestine. Whether this is due to genetic polymorphism or to post translational modification of the enzyme will be the subject of future investigation. It was noted that ALP activity could not be determined after SDS-PAGE electrophoresis as the enzyme activity was inhibited and 2D electrophoresis was not appropriate to investigate the ALP isozymes. The characterisation of bovine nasal ALP will enable the study of the pathophysiological responses of the nasal epithelium to respiratory diseases and may lead to biomarkers of use in diagnosis and monitoring.

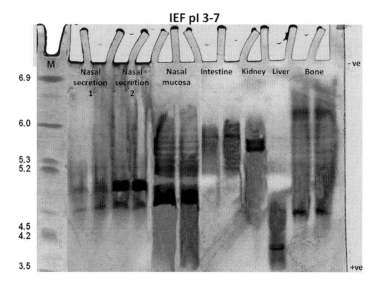

Figure 1. ALP isozyme bands of bovine nasal secretion and various tissue extracts separated on a Invitrogen Novex® pH 3-7 gradient IEF Gel. Lane M show the Coomassie stained IEF standard protein marker, the cathode is indicated by (-ve) and anode by (+ve).

Aknowledgement

The Malaysian Ministry of Higher Education is thanked for funding MFG.

Reference

Ghazali, M.F., Jonsson, N.N., Burchmore, R.J.S. and Eckersall, P.D., 2012. Biochemical and proteomic investigation of bovine nasal secretion. Proceedings of the 3rd Managing Committee Meeting and 2nd Meeting of Working Groups 1, 2 & 3 of COST Action FA1002 2:63-66.

Part III
Infectious diseases

PilE4 may contributes in the adhesion of *Francisella* to brain microvascular endothelial cells

Elena Bencurova[1], Patrik Mlynarcik[1], Lucia Pulzova[1,2], Andrej Kovac[2] and Mangesh R. Bhide[1,2]
[1]*University of veterinary medicine and pharmacy in Košice, Slovakia; mangeshbide@me.com*
[2]*Institute of Neuroimmunology of Slovak Academy of Sciences, Bratislava, Slovakia*

Objectives

Francisella tularensis (*Ft*), a small Gram-negative facultative intracellular bacterium, is the causative agent of tularemia. Disease is transmitted to human and animals mostly by vectors such as ticks, flies and mosquitoes. Tularemia is endemic in many parts of the northern hemisphere and has been detected in over 250 animals. *Ft* can invade many organs in the host body, such a liver, eyes, lung and central nervous system. Neural form of tularemia is rare, but often ends with fatal consequences (Gangat, 2007; van de Beek *et al.*, 2007). One of the crucial steps in CNS invasion is the crossing of blood-brain barrier (BBB), while, bacterial translocation initiates with their transient adhesion on brain microvascular endothelial cells (BMECs). Present study is aimed at investigation of the molecules responsible for adhesion of *Ft* to BMECs. These molecules could be the important candidates in the development of prophylactic drugs against meningitides caused by *Francisella*.

Material and methods

Cultivation of Francisella and BMECs

Francisella strain LVS was cultivated for 4 days at 37 °C on chocolate agar enriched with 1% glucose and 0.1% L-cystein. Bacterial cells were harvested and whole cell lysate was prepared by sonication. BMECs were prepared from Wistar rats as described previously (Pulzova *et al.*, 2009; Veszelka *et al.*, 2007). Forebrains were cleaned of meninges, minced into small pieces and digested with collagenase type II and DNase. Fragments were separated from myelin layer by centrifugation and microvessels were digested with collagenase-dispase and DNase. Microvessel endothelial cell clusters were separated on a 33% continuous Percoll and endothelial cell clusters were then directly plated on collagen type IV coated chamber slide. Cells were cultivated in DMEM supplemented with 20% plasma derived, gentamicine, L-glutamine, heparin, 1 basic fibroblast growth factor and puromycin. Cells were harvested and whole cell lysate was prepared.

Ligand capture assay to search molecules in BMEC:Francisella interface

Francisella-BMECs protein interaction was assessed by ligand capture assay (LCA). Shortly, *Ft* lysate was fractionated by non-reducing SDS-PAGE and proteins were electrotransferred on nitrocellulose membrane. Membrane was blocked in TBS (100 mM Tris-HCl, pH 7.2 and 150 mM NaCl) containing 0.5% bovine serum albumin fraction V (BSA-V) and membrane was hybridized with cell lysate of BMECs or TBS+BSA-V buffer (negative control). Membranes were washed with TTBS (TBS with 0.05% Tween 20) and incubated in protein capture buffer (patent pending). Capture buffer containing proteins of BMEC interacting with *Ft* ligand were collected, desalted and concentrated with MWCO filters (PES-5000, Sartorius, Germany), fractionated by non-reducing SDS PAGE and visualized by silver staining. Protein observed on the gel was excised, trypsin digested and processed for peptide mass fingerprinting (Shevchenko *et al.*, 1996).

Validation of ICAM-1:Ft interaction

His tagged rICAM-1 was over-expressed in *Saccharomyces cerevisiae* strain YPD 501 and purified with metal affinity chromatography. Truncated rICAM-1 was then immobilized on Co^{2+} metal affinity magnetic beads (Bruker-Daltonics, Germany) with gentle agitation for 1 hr. Beads were washed three times and then hybridized with whole cell lysate of *Ft* resuspended in binding buffer. After three washings, protein complex was eluted with elution buffer (all buffers from Bruker-Daltonics, Germany), separated on SDS-PAGE and visualized by coomassie blue staining. Protein band observed on the gel, apart from the rICAM-1, was excised and identified by peptide mass fingerprinting.

Confirmation of ICAM-1:Ft interaction

Pull down assay was performed to confirm interaction between BMECs:PilE4. Shortly, His-tagged rPilE4 (overexpressed in *E. coli* strain SG13009) was immobilized on metal affinity beads (Talon resin, Clontech, USA), hybridized overnight with whole cell lysate of BMECs with constant shaking and washed twice with wash buffer. Protein complex was eluted with 250 mM imidazole and proteins were desalted by Zip-Tips (Millipore, USA). Desalted proteins were mixed with sDHB matrix in TA50 and subjected for MALDI-TOF mass spectrometry.

Assessment of binding site for PilE4 on ICAM-1

Five coding regions corresponding to different domains of ICAM-1 (Ig-like-C2-type domain 1 to 5) were amplified with PCR. Overlap extension PCR (OE-PCR) was performed to fuse PCR products in the expression cassette containing T7 promoter, species independent translation sequence (SITS, which encompasses poly TTTTA region) and start codon at 5', while eGFP, Myc-tag, stop codon and untranslated region at 3'. Expression cassettes were translated *in vitro* (*in-vitro* translation kit, Jena bioscience, Germany) for 2 hrs at 20 °C and kept at 4 °C overnight for maturation of protein. Proper translation and the presence of protein were detected with epifluorescence of tagged GFP.

ALBI assay was performed to identify domain/s of ICAM-1 responsible for binding of PilE4. Domains were immobilized on PVDF membrane and then incubated with His-tagged rPilE4. After washing with TTBS, membrane was incubated with Ni-HRP conjugate diluted at 1:5,000 in 1% BSA-TTBS. Membrane was incubated with chemiluminisence solution (Pierce, USA) and signals were captured on X-ray film.

Results

Ligand capture assay performed to identify protein candidates involved in *Francisella*:BMEC interface revealed one prominent protein (Figure 1 panel A) which was identified as ICAM-1 by peptide mass fingerprinting (6 peptides matched, significance threshold <0.05). This might be the most probably adhesion receptor for *Francisella*. To identify the ligand interacting with ICAM-1, magnetic beads based pull down assay was perform in which two protein bands

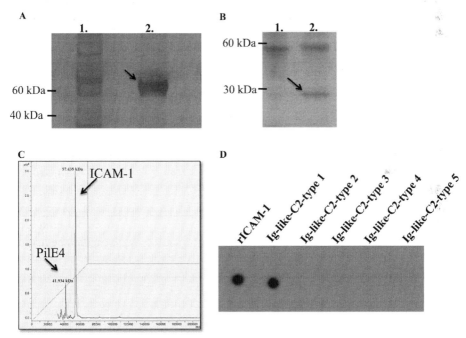

Figure 1. Series of experiments performed to reveal ligand:receptor interaction in Francisella-BMEC interface. Panel A, Ligand capture assay (LCA) performed to identify protein of BMECs interacting with Ft. lane 1, whole cell lysate of Ft; lane 2, protein probably interacting with Ft eluted in elution buffer in LCA. Panel B, magnetic beads based pull down assay to determine Ft ligand responsible to interaction with rICAM-1. Lane 1, rICAM-1 was captured on magnetic beads and eluted, lane 2, rICAM-1 was captured on magnetic beads, hybridized with whole cell lysate of Ft and eluted. Panel C, recombinant PilE4 was bound on affinity beads and incubated protein extracted from rat BMECs. Panel D, Binding site for PilE4 on ICAM-1.

were observed; ~57 kDa protein band corresponding to rICAM-1 and ~35 kDa protein band (Figure 1 panel B) identified later as type IV pili fiber building block protein (PilE4, 6 peptides matched, significance threshold <0.05).

To verify binding ability between PilE4 and ICAM-1, next round of pull down assay was performed in which recombinant form of PilE4 was used. Elute from pull down analyzed by mass spectrometry showed presence of two peaks, 41.9 kDa and 57.4 kDa peaks corresponding to recombinant form of PilE4 (rPilE4) and ICAM-1, respectively (Figure 1 panel C).

ICAM-1 consists of five domains with distinct functions, which provide binding sites for different ligands (Staunton *et al.*, 1990). Deeper analysis performed to identify binding site on ICAM-1 for PilE4 showed clear interaction between rIg-like-C2-type 1 (domain 1) and rPilE4 (Figure 1 panel D).

Conclusion

Ligand:receptor interaction exposed in this study shows that pilin subunit PilE4 and ICAM-1 might be the most probable molecules involved in the initial stages (transient adhesion) of translocation of *Francisella* across BBB; and Ig-like-C2 type 1 domain of ICAM-1 is the binding site for PilE4.

Acknowledgments

Work was performed in collaboration with Dr. E. Chakurkar, at ICAR Goa India, mainly for transfection and in-vivo protein production. Financial support was from APVV-0036-10 and VEGA−1/0054/12.

References

Gangat, N., 2007. Cerebral abscesses complicating tularemia meningitis. Scand J Infect Dis 39: 258-261.

Pulzova, L., Bhide, M.R. and Andrej, K., 2009. Pathogen translocation across the blood-brain barrier. FEMS Immunol Med Microbiol 57: 203-213.

Shevchenko, A., Wilm, M., Vorm, O. and Mann, M., 1996. Mass spectrometric sequencing of proteins silver-stained polyacrylamide gels. Anal Chem 68: 850-858.

Staunton, D.E., Dustin, M.L., Erickson, H.P. and Springer, T.A., 1990. The arrangement of the immunoglobulin-like domains of ICAM-1 and the binding sites for LFA-1 and rhinovirus. Cell 61: 243-254.

Van de Beek, D., Steckelberg, J.M., Marshall, W.F., Kijpittayarit, S. and Wijdicks, E.F., 2007. Tularemia with brain abscesses. Neurology 68: 531.

Veszelka, S., Pasztoi, M., Farkas, A.E., Krizbai, I., Ngo, T.K., Niwa, M., Abraham, C.S. and Deli, M.A., 2007. Pentosan polysulfate protects brain endothelial cells against bacterial lipopolysaccharide-induced damages. Neurochem Int 50: 219-228.

Comparative proteomics analysis of campylobacteriosis in human and swine: a study of intestine epithelial cell response to bacterial infection

Carmen Aguilar[1], Fernando Corrales[2], Angela Moreno[1] and Juan J. Garrido[1]

[1]Grupo de Genómica y Mejora Animal, Departamento de Genética, Universidad de Córdoba, Spain; b42agjuc@uco.es

[2]Unidad de Proteómica, Centro de Investigaciones Médicas Aplicadas, Universidad de Navarra, Pamplona, Spain

Background

Campylobacteriosis remains the most frequently reported zoonotic disease in humans in the EU with 212,064 cases in 2010 (Team, 2012). *Campylobacter* spp. naturally colonizes the intestine of humans and pigs. However, while in humans is the most commonly recognized cause of bacterial gastroenteritis, in swine, due to its typically high prevalence, these bacteria are considered as commensal, making this food-animal in a natural reservoir potentially hazardous for public health (Horrocks *et al.*, 2009). The mechanisms underlying these differences are not well known although could be related to the host immune response.

In this work we use an *in vitro* model to examine the interaction of *Campylobacter jejuni* and *Campylobacter coli* with human and porcine intestinal epithelial cells. Bacterial pathogenesis is a multifactorial process mediated by the adherence of bacteria to host epithelial cells and the response of the host immune system to infection. Proteomics offers excellent possibilities for determining the proteins involved in each of these two biological processes. Therefore, the aim of the present study was to perform a differential proteome analysis to contribute to the knowledge of the differences in *Campylobacter* pathology between human and pigs.

Material and methods

An *in vitro* model of bacterial infection of human and pig intestinal cell lines (INT-407 and IPEC-1, respectively) with swine isolates of *C. jejuni* and *C. coli* was performed (MOI of 100).

After 3h of incubation, cells were washed three times with ice cold PBS and lysated with lysis buffer (Tris 30mM pH 8, Urea 7M, tiourea 2M, CHAPS 4%, ph8.5). After clean up precipitation (Biorad, USA), protein samples were quantified by Bradford's method and solubilized in 2D DIGE sample buffer. Then 50 µg from each sample were labeled with 400 pmol of CyDye (DyeAGNOSTICS, Germany) (Cy3, Cy5 for samples and Cy2 for internal control consisting of a mixture composed by equal amounts of protein from all samples) and incubated for 30 min on ice in the dark. Labeling reaction was stopped by addition of lysine. Samples were loaded onto IPG strips, 24 cm, pH 3–11NL (GE Healthcare, UK), and subjected to isoelectric focusing in IPGphor™ IEF System (GE Healthcare, UK) according to the manufacturer's recommendations. Upon IEF, strips were equilibrated first for 15 min in reducing solution (100 mM Tris–HCl, pH 8.8, 6 M urea, 30% glycerol, 2% SDS and 0.5% DTT) and secondly for 15 min in alkylating solution (100 mM Tris–HCl, pH 8.8, 6 M urea, 30% glycerol, 2% SDS and 4.5% iodoacetamide). For the second dimension, IPG strips were loaded on top of 12.5% polyacrylamide gels and run (1 W/gel) for 12-14 h using Ettan-DALTsix electrophoresis unit (GE Healthcare, UK).

Afterwards, 2D gels were scanned using a Typhoon™ Trio Imager (GE Healthcare, UK). Image analysis was performed using DeCyder 6.5 software (GE Healthcare, UK) as described in the user's manual. Three independent experiments were performed for each experimental setup. Human and porcine experiments were analyzed separately. Differential expressed spots were considered for MS analysis when the fold change was larger than 1.1 and the P-value after t-test was below 0.05. Spots differentially represented were excised manually from DIGE gel and in-gel tryptic digestion was performed with 12.5 ng/µl trypsin (in 50 mM ammonium bicarbonate) for 12 h at 37 °C. The resulting peptides were extracted with 5% formic acid, 50% acetonitrile. After speed-vac concentration, samples were analyzed by MS.

NanoLC-ESI-MS/MS analysis was performed. Data processing was performed with MassLynx 4.0. Database searching was done with two independent search engines: Phenyx 2.6 (GeneBio) and Mascot Server 2.2 (Matrix Science) against UniprotKB/Swiss-Prot Release 51.6 with 257964 entries. Parameters used in Mascot searches were: enzyme, trypsin; variable modifications, carbamidomethylation of cysteine and oxidation of methionine; maximum missed cleavages, 1; peptide mass tolerance settings/windows was 50 ppm; product mass tolerance, 0.1 Da. Probability p of random matches was set to the default value of 0.05.

Bioinformatics functional analysis was carried out with Ingenuity Pathway Analysis (IPA, Ingenuity Systems, www.ingenuity.com) in order to translate proteomic data from each host to functional biological information.

Results and discussion

In this work we use a comparative proteome analysis approach to decipher the molecular basis of the divergence in the host response against *Campylobacter* that could explain the commensal or pathogenic relationship between bacteria and their host. The results showed a greater number of differentially expressed proteins (DEP) in human epithelial cells than in porcine cells (Figure 1A). In addition, in both cellular types the response against *C. jejuni* was higher than in *C. coli* samples. In particular, 22 differential expressed spots (corresponding to 19 unique proteins) were detected in INT407 cells after *C. jejuni* infection and 2 in the *C. coli* infection. On the other hand, in porcine cell line the number of DEP were 4 and 3, respectively.

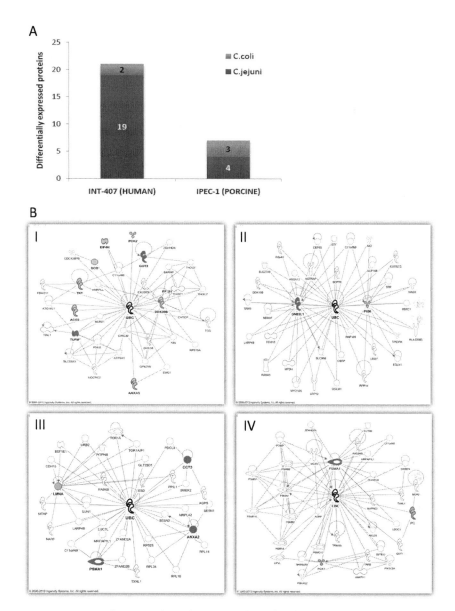

Figure 1. Comparative functional analysis of epithelial cells response against Campylobacter *spp. (A) Number of Differentially Expressed Proteins (DEP) in human (INT407) and porcine epithelial cell (IPEC1) regulated after* Campylobacter *spp. interaction, (B) Functional analysis by IPA®: Networks show relationships among immune response-related proteins in human cells (I and II, agaisnt* C. jejuni *and* C. coli, *respectively) and in porcine cells (III and IV,* C. jejuni *and* C. coli, *respectively). The color of the node indicates whether proteins were up (dark grey) or down (light grey) regulated. Nodes are displayed using various shapes that represent the functional class of the protein.*

Interestingly, most of the DEP identified in human cell line were downregulated after *Campylobacter* infection, some of them with a considerable fold change, as antimicrobial peptide dermcidin, T-complex protein 1 subunit zeta, spliceosome RNA helicase or eukaryotic translation initiation factor 4H (-2.32, -2.6, -3.95 and -3.61 fold change, respectively) in contrast with porcine proteins which were mostly overexpressed. We used Ingenuity Pathway Analysis (IPA®) for functional study of the protein list in each system. This software showed that human DEP were involved in immune-related functions, such an immunological disease, inflammatory disease and inflammatory response functions after *Campylobacter* incubation. Instead, porcine DEP were related with gastrointestinal disease. A range of proteins involved in a very extensive spectrum of cellular pathways was detected in human cells, including the tricarboxylic acid cycle, pentose phosphate pathway or aspartate biosynthesis. This coupled with the fact that a high percent of human identified proteins was localized in mitochondria (mitochondrial proteins) could be related with alterations of the oxidative stress and nitric oxide production detected in human cells after *Campylobacter* infection.

Furthermore, even though no changes were observed in the inflammatory response in porcine cells, apoptosis signaling pathway was altered, indicating some immune regulation in swine cells as well.

Finally, despite network analysis integrated different molecules in porcine and human host, ubiquitin C was revealed as important protein in both networks (Figure 1B). Curiously, the expression pattern of these proteins was opposite (overexpressed in pig and sub expressed in human) showing antagonistic tendency in their immune response, which could be on the basis of the different behavior of *Campylobacter* in both hosts.

Acknowledgements

Work was performed in collaboration with Dr. Fernando Corrales group in a short-term stay supported by EADGENE Funding (European Animal Disease Genomics Network of Excellence for Animal Health and Food Safety).

References

Horrocks, S.M., Anderson, R.C., Nisbet, D.J. and Ricke, S.C., 2009. incidence and ecology of *Campylobacter jejuni* and *coli* in animals. Anaerobe 15: 18-25.

Team, E.E., 2012. The European Union summary report on trends and sources of zoonoses, zoonotic agents and food-borne outbreaks in 2010. Eurosurveillance 17: 21-21.

Identification of amino acid residues of OspA of *Borrelia* involved in binding to CD40 receptor

Patrik Mlynarcik[1], Lucia Pulzova[1,2], Stanislav Hresko[1], Elena Bencurova[1], Saskia Dolinska[1], Andrej Kovac[2], Miguel A. Dominguez[3], Juan J. Garrido[3] and Mangesh R. Bhide[1,2]
[1]*Laboratory of Biomedical Microbiology and Immunology, Department of microbiology and immunology, The University of Veterinary Medicine and Pharmacy in Kosice, Kosice, Slovakia;* patrik.mlynarcik@gmail.com
[2]*Institute of Neuroimmunology of Slovak Academy of Sciences, Bratislava, Slovakia*
[3]*Grupo de Genómica y Mejora Animal, Departamento de Genética, Universidad de Córdoba, Córdoba, Spain*

Objectives

In a previous study we have shown, that the OspA of *Borrelia garinii* (neuroinvasive strain SKT-7.1) is crucial for its transient tethering to the rat brain microvascular endothelial cells (RBMEC) via 31.8 kDa receptor CD40. Transient tethering is then followed by stationary adhesion in which ICAM-1 or VCAM-1 might be the potential molecules. Both transient and stationary adhesions are necessary steps in the traversal of *Borrelia* across the blood-brain barrier (BBB). Understanding the basic mechanisms involved in crossing the BBB by *Borrelia* will help to develop novel therapeutic approaches against neuroborreliosis.

To this background, objectives were set to (1) predict and identify the part of OspA interacting with CD40 molecule and (2) identify amino acids in the OspA responsible for CD40 binding.

Material and methods

In silico mapping of the OspA region interacting with CD40 molecule

The amino acid sequence of OspA was subjected to mapping of endothelium binding sites based on database search (Uniprot, SMART) and data mining (Comstock *et al.*, 1993; McGrath *et al.*, 1995; Pal *et al.*, 2000). Moreover, nucleotide sequences of OspA of SKT-7.1 and SKT-2 (GenBank no. GU906888 and GU320003, respectively) were *in silico* translated and amino acid sequences were aligned using Clustal W (Geneious Pro, Biomatters).

Preparation of truncated OspA fragments

Four N-terminal truncated variants of OspA (1) OspA-full (18-249), (2) TGE 1,2 (18-132), (3) HUVE (127-205), (4) TGE 3 (204-261), were prepared to find interacting part on OspA for CD40 (amino acids numbering based on GenBank no. GU906888). The fragments were cloned in pQE-30 UA-GFP (Qiagen) vector and proteins were overexpressed in *E. coli* strain SG13009.

Further truncation of HUVE fragment and far-western blotting

The HUVE fragment was further truncated into (1) HUVE(127-155), (2) HUVE(154-205), (3) HUVE(127-171) and (4) HUVE(127-185). Proteins extracted from RBMEC were separated by SDS-PAGE and electro-transferred on nitrocellulose membrane. Membrane was hybridized with truncated OspA, followed by incubation with HisProbe-HRP conjugate (Pierce, USA). Presence of recombinant protein was then visualized on X-ray film using the ECL (chemiluminescence) system (Pierce, USA).

Pull-down assay of OspA fragments

Shortly, His-tagged proteins were bound on Talon beads (Clontech, USA) and then incubated with cleared whole cell lysate prepared from primary cultures of RBMEC. The complex of interacting proteins was eluted and subjected to MALDI-TOF analysis.

Preparation of OspA mutants

Twenty amino acids varied in HUVE fragment of SKT-7.1 and SKT-2 were subjected to site-directed mutagenesis (Table 1). 20 mutants of OspA of SKT-7.1 in pQE-30 UA-GFP vector were prepared with QuikChange Lightning Multi Site-Directed Mutagenesis Kit (Stratagene, USA). Mutant plasmids with correct point-mutation were purified using GenElute Plasmid Midi-Prep Kit (Sigma-Aldrich) and transformed into SG13009 cells (Qiagen). Sequences containing mutations were submitted to GenBank repository: JX889248-JX889267. The OspA mutant proteins were overexpressed and purified by Talon metal chelate affinity chromatography (Clontech).

Table 1. Overview of mutation sites in OspA.

Mutation number	Amino acid substitution	Mutation number	Amino acid substitution
1	Thr132Val	11	Gly-insertion175Ala
2	Thr136Ile	12	Lys179Val
3	Val138Thr	13	Thr181Lys
4	Asn141Asp	14	Val186Thr
5	Asp149Gly	15	Leu192Ser
6	Asp164Gly	16	Ile197Val
7	Phe165Tyr	17	Thr198Ser
8	Thr166Val	18	Ala200Glu
9	Ala172Thr	19	Asp202Asn
10	Asp174Glu	20	Ser204Thr

Preparation of r-CD40 and CD40:OspA mutants interaction

PCR amplified CD40 fragment (28-192 AA) was ligated into *pLEXSY_I-blecherry3* (amino acids numbering based on GenBank no. BC097949). Expression cassette was amplified with long PCR using the LongRange PCR Kit (QIAGEN), purified with QIAquick Gel extraction kit (Qiagen) and transformed into LEXSY-T7-TR strain (Jena Biosciences, Germany) by electroporation. T7-TR cell with overexpressed Myc-tagged CD40 was lysed and proteins were resolved by SDS-PAGE, transferred to nitrocellulose membrane and incubated with OspA mutant proteins prepared above. The interaction was detected by HisProbe-HRP conjugate and ECL substrate.

Results and discussion

Previously, it was confirmed that OspA of SKT-7.1 (neuroinvasive strain of *B. garinii*), but not from SKT-2 (non-neuroinvasive strain), interacts with rat CD40 receptor (Pulzova *et al.*, 2011). In order to determine which part of the OspA takes part in CD40:OspA interface, series of *in silico* analysis were performed, which revealed that OspA contains four endothelial cell binding sites. Further, comparative sequence analysis of OspA from SKT-2 and SKT-7.1 showed 49 amino acid variations (details not shown), that may affect the affinity of OspA to CD40.

Hybridization experiments performed with truncated OspA and RBMEC protein showed that only HUVE fragment was able to interact. These results indicate that CD40 binding site on OspA is located in this region. Further truncation of HUVE subunit completely abolished the ability of truncated protein to bind to CD40 (Figure 1, Panel A). Taken together, these results indicate that a 79-amino acid region between amino acids 127 and 205 most likely forms the CD40-receptor binding site.

Site-directed mutagenesis was performed to determine importance of amino acids that varied among OspA of SKT-7.1 and SKT-2 in the binding to CD40 molecule. Interestingly, it was found that all OspA mutants, except mutations Asp149Gly, Phe165Tyr, Ala172Thr, Val186Thr and Leu192Ser, readily bound CD40. Results indicate that these five amino acid residues principally contribute to CD40 interaction or are essential for juxtaposition of OspA:CD40 (Figure 1, Panel B).

Our results suggest that the string of amino acids located in the HUVE part constitutes an essential part of the CD40-receptor binding site. Moreover, the extracellular part of the OspA consisting of these five amino acids may serve as a vaccine target and evoke production of protective antibodies against infection.

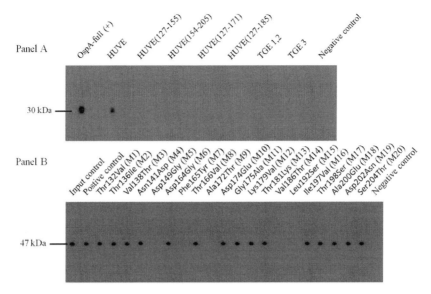

Figure 1. Far-western blots showing CD40-binding ability of different OspA constructs. Panel A – presents binding of truncated OspA with ~30 kDa protein of RBMEC. Negative control – membrane was incubated only with whole cell lysate of E. coli containing no OspA expression cassette. Panel B – shows binding of OspA mutants with CD40. Input control – r-CD40 immobilized on membrane was detected with Anti-Myc-HRP antibody; positive control – wild type OspA from SKT-7.1 was hybridized with r-CD40 on membrane; negative control – any non-specific binding of HisProbe-HRP conjugate to r-CD40 was ruled out by hybridizing the HisProbe-HRP with CD40 only.

Acknowledgements

Work was performed in collaboration with Dr. E. Chakurkar, at ICAR Goa India, mainly for transfection and *in vivo* protein production. Financial support was from APVV-0036-10, VEGA–1/0054/12, 2/0193/11. PM received SAIA-NSP funding for short term stay at JJG's Lab for site-directed mutagenesis experiments.

References

Comstock, L.E., Fikrig, E., Shoberg, R.J., Flavell, R.A. and Thomas, D.D., 1993. A monoclonal antibody to OspA inhibits association of *Borrelia burgdorferi* with human endothelial cells. Infect Immun 61: 423-431.

McGrath, B.C., Dunn, J.J., Gorgone, G., Guttman, D., Dykhuizen, D. and Luft, B.J., 1995. Identification of an immunologically important hypervariable domain of major outer surface protein A of *Borrelia burgdorferi*. Infect Immun 63: 1356-1361.

Pal, U., de Silva, A.M., Montgomery, R.R., Fish, D., Anguita, J., Anderson, J.F., Lobet, Y. and Fikrig, E., 2000. Attachment of *Borrelia burgdorferi* within *Ixodes scapularis* mediated by outer surface protein A. J Clin Invest 106: 561-569.

Pulzova, L., Kovac, A., Mucha, R., Mlynarcik, P., Bencurova, E., Madar, M., Novak, M. and Bhide, M., 2011. OspA-CD40 dyad: ligand-receptor interaction in the translocation of neuroinvasive *Borrelia* across the blood-brain barrier. Sci Rep 1: 86.

Omics approaches to study the Rickettsia *Ehrlichia ruminantium*: towards improved knowledge on Heartwater disease

Isabel Marcelino[1,2], Miguel Ventosa[1,2], Ludovic Pruneau[3*], Elisabete Pires[2], Damien F. Meyer[3], André M. de Almeida[2,4], Bernard Mari[5], Thierry Lefrançois[6], Ana V. Coelho[2] and Nathalie Vachiéry[3]*

[1]Instituto de Biotecnologia Experimental e Tecnológica (IBET), Apartado 12, 2780-901 Oeiras, Portugal*

[2]Instituto de Tecnologia Química e Biológica, Universidade Nova de Lisboa (ITQB), Av. da República, 2780-157 Oeiras, Portugal; marcelino@itqb.unl.pt

[3]Centre de coopération Internationale en Recherche Agronomique pour le Développement (CIRAD), UMR CMAEE, 97170 Petit-Bourg, Guadeloupe, FWI

[4]Instituto de Investigação Científica Tropical, Centro de Veterinária e Zootecnia (IICT/CVZ) & Centro Interdisciplinar de Investigação em Sanidade Animal (CIISA)), Fac. Med. Veterinária, Av. Univ. Técnica, 1300-477 Lisboa, Portugal

[5]Institut de Pharmacologie Moléculaire et Cellulaire, UMR 6097, CNRS-Université de Nice Sophia Antipolis, Sophia Antipolis, 06560, France

[6]Centre de coopération Internationale en Recherche Agronomique pour le Développement (CIRAD), UMR CMAEE, 34398 Montpellier, France

*Both authors equally contributed for this work

Objectives

The Rickettsia *Ehrlichia ruminantium* is an obligate intracellular bacterium that causes Heartwater, a fatal tick-borne disease in ruminants, posing important economical constraints to livestock production in sub-Saharan Africa and in some Caribbean islands, where it threatens the American mainland. Several vaccine candidates (inactivated, attenuated and recombinant) are under evaluation, but the development of a fully effective vaccine has been hindered by the high antigenic diversity between strains and the lack of knowledge on *E. ruminantium* biology.

In host endothelial cells, *E. ruminantium* has a biphasic life cycle: infection starts with adhesion and invasion of infectious particles called elementary bodies (EB). Once inside the cell, EBs differentiate into reticulate bodies (RB) which divide within a vacuole called inclusion. After several days, RBs re-differentiate into EBs which are then released and start a new infectious cycle (Jongejan *et al.*, 1991; Marcelino *et al.*, 2005). To successfully survive and multiply in these different environments, *E. ruminantium* requires life cycle stage-specific adaptation. Although three genome sequences are currently available (Collins *et al.*, 2005; Frutos *et al.*, 2007), *E. ruminantium* pathogenesis is, so far, poorly understood.

In this work we used global Omics approaches such as transcriptomics and proteomics to identify genes and proteins differentially expressed (1) in EBs and RBs and (2) between virulent and attenuated bacterial strains, both currently being tested as inactivated and attenuated vaccines, respectively.

Material and methods

E. ruminantium Gardel strain isolated in Guadeloupe (FWI) was used throughout these studies. Virulent (ERGvir) and attenuated (ERGatt) phenotypes are routinely cultivated *in vitro* in bovine aortic endothelial cells, BAE (Marcelino *et al.*, 2005; Pruneau *et al.*, 2012). ERGatt derives from ERGvir after over 200 laboratory passages in host cell. *E. ruminantium*-infected BAE cells containing colonies of reticulate bodies were harvested at the mid-exponential phase of the bacterial developmental phase while EBs were harvested at 80% cell lysis. For transcriptomics studies, at the time of harvesting, *E. ruminantium* samples were immediately treated with TRIzol reagent (Invitrogen, France) for total RNA extraction. Bacterial cDNA without or with selective capture of *E. ruminantium* transcripts (Emboule *et al.*, 2009) were hybridized on *E. ruminantium* custom-made microarrays (8 × 60 k, Agilent, France). Hybridization and statistical analysis were performed as described in Pruneau *et al.* (2012). Genes differentially expressed between virulent and attenuated strains and between times post-infection will be annotated using bioinformatic tools.

For proteomics analyses, EB samples were purified by a multistep centrifugation process (Marcelino *et al.*, 2007). Total *E. ruminantium* protein extracts were obtained using sonication, quantified by 2D Quant kit (GE, Uppsala, Sweden) and analyzed using 1DE, 2DE and DIGE gels. Protein bands were excised, digested and submitted to nano-LC-MALDI-TOF/TOF-MS analysis. MS/MS Spectrum acquired in ABSciex® MALDI TOF-TOF 4800 were submitted to database search using Peaks® and combining 3 search engines MASCOT, X! Tandem and Peaks for protein identification. To improve the functional annotations of hypothetical proteins (which constitute 30% of the genome), we performed a manual re-annotation of proteins with unknown functions.

Results and discussion

In EBs, transcriptomics and proteomics analyses revealed the presence of several genes and proteins related to energy, aminoacid and general metabolism, to protein turnover, and to chaperones and survival (Marcelino *et al.*, 2012; Pruneau *et al.*, 2012). This indicates that infectious extracellular EBs could be metabolically active and able to protect themselves against harsh conditions outside the host cells. When comparing ERGvir and ERGatt, numerous genes and proteins were found to be common. 2DE gels also revealed that, as previously observed (Marcelino *et al.*, 2012), both phenotypes have many proteins expressed as isoforms, again indicating that post-translational modifications can be important to *E. ruminantium* biology. Several genes and proteins were also found to be differentially expressed between ERGvir and

ERGatt. For instance, the MAP1-family proteins and genes were found to be overexpressed in the virulent strain. Overexpression of non-protective MAP1-familly proteins on the bacteria outer membrane could be used by *E. ruminantium* as a strategy to mask protective antigens, hereby delaying a protective immune response in infected animals. Oppositely, genes involved in energy production, nucleotide transport, replication, lipid metabolism were found to be overexpressed on ERGatt strain. This could be related to its well-fitted metabolism and replication in endothelial cells. Interestingly, four major protein spots corresponding to a porin were exclusively found in the attenuated phenotype and correspond to the highest differential expression level on microarrays analyses.

Transcriptomics studies on RBs from the virulent strain showed an over-expression of genes related to energy production and conversion, co-enzyme transport and metabolism, replication, recombination and DNA repair functional categories (Pruneau *et al.*, 2012). Preliminary proteomics results on similar samples indicated that the most abundant proteins detected are related to bacterial cell division (FtsA, DapE and XerC), energy production (AtpD and F), metabolism of pryruvate, amino acids, nucleic acid and protein synthesis and processing. Several outer membrane proteins were also detected. These results are in accordance with the active division phase of *E. ruminantium*.

Conclusion and future prospects

Globally, the work presented herein provides new insights on the genes, proteins and regulatory pathways underlying *E. ruminantium* biology. Indeed, these results allowed the characterization of non-infectious metabolically-active RBs and the extracellular infectious EBs and to find genes and proteins associated to virulent and attenuated phenotypes. This knowledge will be useful to identify (1) biomarkers for RB and EBs, (2) virulence/attenuation mechanisms, and (3) new antigens to develop improved vaccines against heartwater.

Acknowledgements

Authors acknowledge funding from the EU projects COST action FA-1002 (for networking opportunities) and FEDER 2007-2013 (FED 1/1.4-30305) and also the project ER-TRANSPROT (PTDC/CVT/114118/2009), Post-doc grant SFRH/ BPD/ 45978/ 2008 (I. Marcelino) and the Ciência 2007 research contract (AM Almeida), all by Fundação para a Ciência e a Tecnologia (Lisboa, Portugal).

References

Collins, N.E., Liebenberg, J., de Villiers, E.P., Brayton, K.A., Louw, E., Pretorius, A., Faber, F.E., van Heerden, H., Josemans, A., van Kleef, M., Steyn, H.C., van Strijp, M.F., Zweygarth, E., Jongejan, F., Maillard, J.C., Berthier, D., Botha, M., Joubert, F., Corton, C.H., Thomson, N.R., Allsopp, M.T. and Allsopp, B.A., 2005. The genome of the heartwater agent *Ehrlichia ruminantium* contains multiple tandem repeats of actively variable copy number. Proc Natl Acad Sci U S A 102: 838-843.

Emboule, L., Daigle, F., Meyer, D.F., Mari, B., Pinarello, V., Sheikboudou, C., Magnone, V., Frutos, R., Viari, A., Barbry, P., Martinez, D., Lefrancois, T. and Vachiery, N., 2009. Innovative approach for transcriptomic analysis of obligate intracellular pathogen: selective capture of transcribed sequences of *Ehrlichia ruminantium*. BMC Mol Biol 10: 111.

Frutos, R., Viari, A., Vachiery, N., Boyer, F. and Martinez, D., 2007. *Ehrlichia ruminantium*: genomic and evolutionary features. Trends Parasitol 23: 414-419.

Jongejan, F., Zandbergen, T.A., van de Wiel, P.A., de Groot, M. and Uilenberg, G., 1991. The tick-borne rickettsia *Cowdria ruminantium* has a *Chlamydia*-like developmental cycle. Onderstepoort J Vet Res 58: 227-237.

Marcelino, I., de Almeida, A.M., Brito, C., Meyer, D.F., Barreto, M., Sheikboudou, C., Franco, C.F., Martinez, D., Lefrancois, T., Vachiery, N., Carrondo, M.J., Coelho, A.V. and Alves, P.M., 2012. Proteomic analyses of *Ehrlichia ruminantium* highlight differential expression of MAP1-family proteins. Vet Microbiol 156: 305-314.

Marcelino, I., Vachiery, N., Amaral, A.I., Roldao, A., Lefrancois, T., Carrondo, M.J., Alves, P.M. and Martinez, D., 2007. Effect of the purification process and the storage conditions on the efficacy of an inactivated vaccine against heartwater. Vaccine 25: 4903-4913.

Marcelino, I., Verissimo, C., Sousa, M.F., Carrondo, M.J. and Alves, P.M., 2005. Characterization of Ehrlichia ruminantium replication and release kinetics in endothelial cell cultures. Vet Microbiol 110: 87-96.

Pruneau, L., Emboule, L., Gely, P., Marcelino, I., Mari, B., Pinarello, V., Sheikboudou, C., Martinez, D., Daigle, F., Lefrancois, T., Meyer, D.F. and Vachiery, N., 2012. Global gene expression profiling of *Ehrlichia ruminantium* at different stages of development. FEMS Immunol Med Microbiol 64: 66-73.

Detection of *Salmonella* antigens expressed in swine gut by hydrophobic antigen tissue triton extraction (HATTREX)

Rodrigo P. Martins[1], Ruben B. Santiago[2], Angela Moreno[1], Juan J. Garrido[1] and Jarlath E. Nally[2]
[1]*Grupo de Genómica y Mejora Animal, Departamento de Genética, Universidad de Córdoba, Córdoba, Spain; ge1gapaj@uco.es*
[2]*School of Veterinary Medicine, UCD, Dublin, Ireland*

Background and objectives

The outcome of an infection is determined by a delicate balance between pathogen virulence and the ability of host to mount immune response. In this context, approaches that enable the study of mechanisms carried out by pathogens *in vivo* represent fundamental tools for a comprehensive view of infectious processes.

Hydrophobic Antigen Tissue Triton Extraction (HATTREX) is an innovative method that makes possible the characterization of the bacterial membrane proteome during infection (Crother and Nally, 2008). In light of this, we aimed in this study to check the feasibility of HATTREX as an approach to explore the membrane proteome expressed by *Salmonella enterica* serovar Typhimurium (herein, *S. typhimurium*) in ileum of experimentally infected pigs.

Material and methods

S. typhimruium strain and tissue samples employed in this report proceed from an experimental infection described by Collado-Romero *et al.* (2010). Briefly, eight crossbred piglets of approximately four weeks of age, confirmed to be fecal-negative for *Salmonella*, were randomly allocated to control or infected groups (four animals each), being control pigs necropsied 2 h before the experimental infection. Animals belonging to the infected group were orally challenged with 10^8 cfu of *S. typhimurium* phagetype DT104 and euthanized at 2 days post-infection. Then, ileum samples were collected from all piglets and immediately frozen in liquid nitrogen.

For HATTREX analysis of pathogen proteome *in vitro*, *S. typhimurium* was grown in TSB to logarithmic phase (OD_{600} of 0.70). After two washes with TE (10mM Tris-CL, 1mM EDTA, pH 8.0), bacterial cell pellets were processed employing the protocol described by Crother and Nally (2008). The same protocol was used for the enrichment of *Salmonella* hydrophobic protein from a pool of infected ileum and control samples. Subsequently, the obtained protein fractions were routinely separated by 1D or 2D electrophoresis for Sypro Ruby staining and immunoblot assays employing specific rabbit antiserums against *S. typhimurium* (anti-somatic antigen and anti-flagellin) and sera from naturally infected pig.

Results and discussion

HATTREX technique was initially developed for analysis of spirochetes proteins expressed *in vivo*. For this reason, the first part of the work consisted in the standardization of the procedure for *Salmonella* cultures and infected tissue. Adaptations of the published protocol included an increase of Triton X-114 concentration (from 1 to 2.5%) and incubation time at 37 °C (from 10 to 30 minutes) for phase separation of lysates.

Firstly, we confirmed by 1D electrophoresis that HATTREX could be effectively employed to produce phase partitioning of protein lysates from *S. typhimurium* cultures. Considering this, we executed this method to study of the membrane proteome of *S. typhimurium* found in ileum of infected pigs. Hydrophobic proteins could be enriched from both control and infected samples and immunoblots with anti-*Salmonella* specific rabbit antiserums validated that isolated proteins corresponded to pathogen proteins (Figure 1A). Curiously, anti-*Salmonella* somatic antigen antiserum was observed to be cross-reactive with some protein from other organism, since labelling was observed in uninfected animals. However, labelling of extra bands detected exclusively in the infected group suggested the isolation of *Salmonella* proteins expressed *in vivo*. Furthermore, immunoblot analysis employing anti-*Salmonella* flagellin revealed no reaction in control pool and a similar labelling pattern for positive control (*Salmonella* whole cell lysate) and infected samples.

Specific antiserums employed in the initial assays were produced by rabbit immunisations with fixed *S. typhimurium* grown *in vitro*. For this reason, this approach could prevent detection of pathogen proteins expressed only *in vivo*. To address this issue, we screened detergent rich fraction by immunoblots, using the serum of a pig naturally infected with *S. typhimurium*. As illustrated in Figure 1B, isolation of pathogen proteins from infected pig ileum was again demonstrated. Moreover, some spots were also labelled in uninfected control. Since control samples proceed from *Salmonella* free pigs, this result could be attributed to cross-reactions of the employed serum against proteins from host or other bacteria that are naturally present in ileum.

Finally, in order to clarify the results obtained by 1-D electrophoresis, the previous experiment was repeated using hydrophobic fractions from *in vitro S. typhimurium* and ileum (control and infected samples), separated by 2-D electrophoresis. As depicted in Figure 1C-E, some spots could be equivalently detected in both infected tissue and *in vitro S. typhimurium* proteomes. Besides, the distinct labelling pattern observed for uninfected ileum reinforces the hypothesis that labelling observed in this sample is a consequence of cross-detection of proteins from organisms other than *Salmonella*.

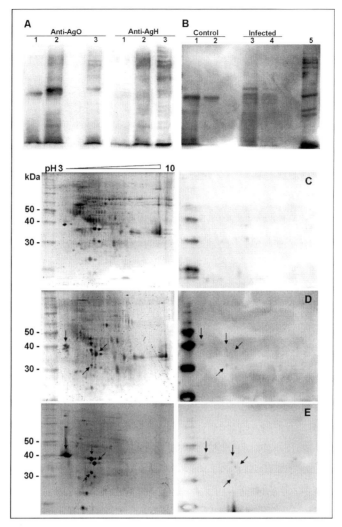

Figure 1. Analysis of S. typhimurium *membrane proteome expressed* in vitro *and in ileum of infected pigs by HATTREX. A: Immunoblot analysis of hydrophobic protein from uninfected controls (1) and infected ileum (2) employing rabbit serums anti-*Salmonella *somatic antigen (anti-AgO) and anti-*S. typhimurium *flagellin (anti-AgH).* S. typhimurium *whole cell lysate (3) was used as positive control. B: Hydrophobic protein from infected ileum and controls was immunolabelled using serum from* Salmonella *naturally infected pig. A total of 7.5 ug of protein were loaded in lanes 1 and 3. In lanes 2 and 4, 1.5 ug of protein was analysed.* S. typhimurium *whole cell lysate (5) was used as positive control. For A and B, arrows highlight differences between infected and control groups. C-E: Sypro Ruby staining (left) and immunoblot analysis (right) of hydrophobic protein separated by 2-D electrophoresis, from ileum of controls (C) and infected (D) pigs and* S. typhimurium *grown* in vitro *(E). Serum from* Salmonella *naturally infected pig was used in this assay. Arrows show spots detected in proteomes of both infected tissue and bacteria grown* in vitro.

Together, these results attest that HATTREX method was able to separate and detect hydrophobic protein expressed by *S. typhimurium* in the gut of infected swine. To our knowledge, this preliminary work is the first trial of exploring the proteome expressed by this pathogen during porcine salmonellosis and represents a base for further research aiming to gain an insight into infection from pathogen perspective.

Acknowledgements

Work was performed during a COST Short Term Scientific Mission (STSM) at the School of Veterinary Medicine, UCD, Dublin, Ireland, in collaboration with Professor J.E. Nally. Financial support was provided by COST action FA1002 [ECOST-STSM-FA1002-011012-019254]

References

Collado-Romero M, Arce C, Ramírez-Boo M, Carvajal A and Garrido JJ., 2012. Quantitative analysis of the immune response upon *Salmonella typhimurium* infection along the porcine intestinal gut. Vet Res 41: 23.

Crother TR, Nally JE., 2008. Analysis of bacterial membrane proteins produced during mammalian infection using hydrophobic antigen tissue triton extraction (HATTREX). Curr Protoc Microbiol 12.1.1-12.1.5.

New approaches in the diagnosis and treatment of endotoxemia in calves: proteomic investigation and effects of intravenous choline administration

Meriç Kocaturk[1], Mehmet Cansev[2], Ahmet Tarik Baykal[3], Orcun Hacariz[3], İbrahim Hatipoglu[3], Oya Eralp Inan[1], Zeki Yilmaz[1] and Ismail Hakki Ulus[4]

[1]*Uludag University, Veterinary Teaching Hospital, Internal Medicine Department, 16059, Gorukle, Bursa, Turkey; merick@uludag.edu.tr*
[2]*Uludag University,Faculty of Medicine, Pharmacology Department, Bursa, Turkey*
[3]*TUBITAK Marmara Research Center, Genetic Engineering and Biotechnology Institute, Kocaeli, Turkey*
[4]*Acıbadem University, Faculty of Medicine, Department of Pharmacology, Maltepe, İstanbul, Turkey*

Introduction

Life-threatening diseases like sepsis/endotoxemia commonly result with death despite comprehensive treatment modalities. Studies are underway regarding understanding the complex pathophysiological mechanism of sepsis, describing criteria for early diagnosis and developing new treatment approaches in order to decrease mortality.

Pathogens (microorganism-related molecular structures) initiate consecutive intracellular events in immune, endothelial and neuroendocrine cells in sepsis/endotoxemia cases. Proinflammatory mediators (TNF-α etc.) eradicate invading microorganisms by initiating acute phase reaction and prevent tissue injury occurring due to endotoxins released from Gram-negative bacteria, while anti-inflammatory mediators (IL-4 etc.) adapt the immune system and leukocyte responses to this new condition in order to take these reactions under control. Although in general mediator release and triggering mechanisms in this process are well characterized, not much is known in terms of molecular mediators (proteins) that initiate and maintain the response before and during inflammation. In addition to proteins that comprise a significant portion of plasma, thousands of proteins, levels of which may alter during diseases exist. However, qualitative and quantitative measurement of such proteins requires advanced technological tools such as proteomic analysis. Recent studies reported that a lot of proteins were altered significantly in endotoxemic rats (Robichaud *et al.*, 2009) and humans with sepsis (Soares *et al.*, 2009). Description of such proteins that initiate and maintain the intercellular, intracellular and tissue response will lead to rapid and accurate diagnosis, detailed investigation of multi-organ failure mechanism and building an information database for development of new drugs. Although it is hard to develop accurate diagnostic and prognostic approaches to sepsis/endotoxemia cases by measuring a few proteins in a sample using conventional methods, assay of thousands of proteins in the same sample within a short period of time is an advantage provided by 'proteomic analysis' in order to achieve the goals rapidly.

Choline administration enhances cholinergic neurotransmission by stimulating the synthesis and release of acetylcholine; body's immune responses against endotoxin are regulated by activation of sympathoadrenal and parasympathetic nerve branches. Survival increases after choline treatment which inhibits cytokine mediators (TNF-α) and enhances phospholipid synthesis (Eastin *et al.*, 1997; Parrish *et al.*, 2008; Tracey, 2007). Our previous studies showed that intravenous choline administration inhibits/relieves multi-organ injury (Ilcol *et al.*, 2009), lipid abnormalities (Ilcol *et al.*, 2009) and hemostasis defects (Yilmaz *et al.*, 2006; Yilmaz *et al.*) due to endotoxin in dogs and rats. In addition, we showed that choline has a protective/regulatory role on levels of various proteins such as fibrinogen, antithrombin III, high mobility group box-1 and paraoxanase-1. In the light of our data, we believe that intravenous choline administration might become a new approach in treatment of calf endotoxemia, a major health problem with high mortality despite comprehensive treatment modalities. This study aims to find answers to the following 3 basic questions:

1. Could proteomic analysis be utilized for diagnosis of endotoxemia in calves?
2. Could proteomic analysis be utilized for treatment monitorization?
3. Could calf endotoxemia be treated by choline administration?

Material and methods

For this purpose, a total of 20 healthy Holstein calves, of 4 weeks age and male were divided into four groups: control, choline (C), lipopolysaccharide (LPS), and LPS+C.

Calves in the control group were injected with 0.9% NaCl within 5 min (5 ml, intravenous). Calves in C and LPS+C groups received choline chloride (1 mg/kg, intravenous). Endotoxin (*E. coli* lipopolysaccharide, LPS, 055:B5) was injected (2 µg/kg, intravenous, once) to the calves of LPS and LPS+C groups. Clinical, hematological and serum biochemical analyses were performed before (baseline) and 24 hrs after the treatments. At the same time, serum choline and proteomic analyses were performed, as well. To avoid nonspecific elevations in serum-free choline levels from spontaneous release of choline from phospholipids, free choline was separated immediately from phospholipid-bound choline by extraction. Samples were analyzed for free choline radioenzymatically as described previously (Ulus *et al.*, 1998). In order to analyze differences in protein expression with various treatments at various time points, an advanced method namely LC-MS^E were utilized (Figure 1).

Results

In control and C group, there were not any statistical differences of parameters studied between baseline and at 24 hrs post-treatments. Endotoxemia was characterized by increased temperature, heart rate and hematocrit value, as well as neutrophilic leukocytosis and thrombocytopenia, at 24 hrs after LPS administration. LPS lead to increase in AST ($P<0.01$), GGT ($P<0.05$), and CK enzyme activities ($P<0.001$) at 24 hrs, compared to their baseline. Calves with endotoxemia had increased serum levels of BUN and Mg ($P<0.05$), whereas serum

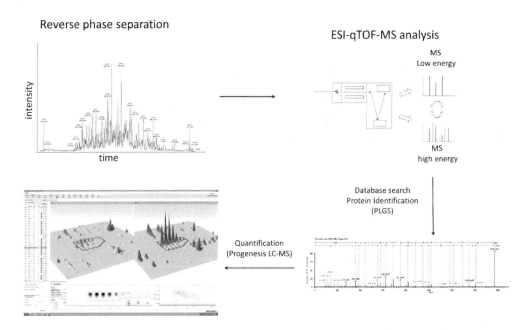

Reverse phase separation

intensity

time

ESI-qTOF-MS analysis

MS
Low energy

MS
high energy

Database search
Protein identification
(PLGS)

Quantification
(Progenesis LC-MS)

Figure 1. Schematic view of LC-MS^E method.

Ca decreased (*P*<0.05), at 24 hrs. ALP, TP, alb, glob and iPhos did not changed significantly during the study in all group. An increase in serum choline concentration in LPS + C group was higher at 24 hrs than that of C and LPS groups. Choline treatment attenuated the clinical and hematological responses to LPS. Choline could minimalize increases in AST, CK, and Mg levels in calves with LPS, as well.

Label-free differential proteome analysis of the serum sample was performed. Data independent acquisition mode was used to collect MS and MS/MS data. A total of 140 proteins were identified across the serum samples. Below is the number of differentially regulated proteins in each sample group (Table 1). Choline, LPS and LPS+Choline treated calf sera at 24 time point was compared to the control samples. Analysis of the samples at time points of 30 min, 1, 4, 24 and 48 hrs is ongoing. Identified proteins will be discussed in the context of the studied system.

Conclusions

This study has potential to present a novel approach in endotoxemia pathophysiology in terms of alteration of functional proteins in response to endotoxin. Intravenous choline administration which possesses a possible therapeutic action in endotoxemia will lead to significant economical benefits by reducing drug costs and decreasing the mortality rate in endotoxemic calves.

Table 1. The number of differentially regulated proteins in each sample group.

Treatment	Number of differentially regulated proteins
Choline	25
LPS	53
LPS+Choline	41

Acknowledgements

This project has been supported by the Scientific and Technological Research Council of Turkey (TOVAG-111O026).

References

Eastin, C.E., McClain, C.J., Lee, E.Y., Bagby, G.J. and Chawla, R.K., 1997. Choline deficiency augments and antibody to tumor necrosis factor-alpha attenuates endotoxin-induced hepatic injury. Alcohol Clin Exp Res 21: 1037-1041.

Ilcol, Y.O., Yilmaz, Z., Cansev, M. and Ulus, I.H., 2009. Choline or CDP-choline alters serum lipid responses to endotoxin in dogs and rats: involvement of the peripheral nicotinic acetylcholine receptors. Shock 32: 286-294.

Parrish, W.R., Rosas-Ballina, M., Gallowitsch-Puerta, M., Ochani, M., Ochani, K., Yang, L.H., Hudson, L., Lin, X., Patel, N., Johnson, S.M., Chavan, S., Goldstein, R.S., Czura, C.J., Miller, E.J., Al-Abed, Y., Tracey, K.J. and Pavlov, V.A., 2008. Modulation of TNF release by choline requires alpha7 subunit nicotinic acetylcholine receptor-mediated signaling. Mol Med 14: 567-574.

Robichaud, S., Lalu, M., Udenberg, T., Schulz, R. and Sawicki, G., 2009. Proteomics analysis of changes in myocardial proteins during endotoxemia. Journal of Proteomics 72: 648-655.

Soares, A.J., Santos, M.F., Trugilho, M.R., Neves-Ferreira, A.G., Perales, J. and Domont, G.B., 2009. Differential proteomics of the plasma of individuals with sepsis caused by *Acinetobacter baumannii*. J Proteomics 73: 267-278.

Tracey, K.J., 2007. Physiology and immunology of the cholinergic antiinflammatory pathway. J Clin Invest 117: 289-296.

Ulus, I.H., Ozyurt, G. and Korfali, E., 1998. Decreased serum choline concentrations in humans after surgery, childbirth, and traumatic head injury. Neurochem Res 23: 727-732.

Yilmaz, Z., Ilcol, Y.O., Torun, S. and Ulus, I.H., 2006. Intravenous administration of choline or cdp-choline improves platelet count and platelet closure times in endotoxin-treated dogs. Shock 25: 73-79.

Yilmaz, Z., Ozarda, Y., Cansev, M., Eralp, O., Kocaturk, M. and Ulus, I.H., Choline or CDP-choline attenuates coagulation abnormalities and prevents the development of acute disseminated intravascular coagulation in dogs during endotoxemia. Blood Coagul Fibrinolysis 21: 339-348.

Changes on bovine aorta endothelial cells (BAE) proteome upon infection with the rickettsia *Ehrlichia ruminantium*

Elisabete Pires[1], Isabel Marcelino[1,2], Nathalie Vachiéry[3], Thierry Lefrançois[4], Gabriel Mazzuchelli[5], Edwin De Pauw[5] and Ana V. Coelho[1]

[1]*Instituto de Tecnologia Química e Biológica, Universidade Nova de Lisboa, Av. da República, 2780-157 Oeiras, Portugal*

[2]*Instituto de Biotécnologia Experimental e Tecnológica, Oeiras, 2780-901 Oeiras, Portugal; marcelino@itqb.unl.pt*

[3]*Centre de coopération Internationale en Recherche Agronomique pour le Développement, UMR CMAEE, F-97170 Petit-Bourg, Guadeloupe, FWI*

[4]*Centre de coopération Internationale en Recherche Agronomique pour le Développement, UMR CMAEE, F-34398 Montpellier, France*

[5]*Laboratory of Mass Spectrometry – GIGA, Proteomics University of Liege, Bat. B6c, B-4000 Liège, Belgium*

Introduction

The rickettsia *Ehrlichia ruminantium* (ER) is an obligatory intracellular bacterium transmitted by *Ambyomma* ticks and the causative agent of heartwater, a fatal tick-borne disease of ruminants in sub-Saharan Africa and in the Caribbean with high economical impact in endemic countries (Allsopp 2010). Several vaccine candidates are being tested, but none has shown to be fully effective. This is due to the high ER antigenic diversity and the lack of knowledge on ER pathogenesis.

In host endothelial cells, ER has a life cycle with two developmental forms: infection starts with the adhesion and invasion of infectious particles called elementary bodies (EB). Once inside the cell, EBs differentiates into non-infectious, metabolically active reticulate bodies (RB) which divide within a vacuole called morula. After several days, RBs re-differentiate into EBs, which are released and start a new infectious cycle. The successful establishment and maintenance of this life cycle must depend on ER capability to hijack specific host cell functions and use host resources for its own need. Despite the recent data on ER transcriptomics (Pruneau *et al.* 2012) and proteomics (Marcelino *et al.* 2012), the mechanisms underlying the impact of ER on host cells remains mostly unknown.

In this work, we analyzed for the first time the effect of ER on the proteome of host bovine aorta endothelial cells (BAE), at the bacteria's mid exponential phase of development (72 hours post-infection (Marcelino *et al.* 2005)), in order to: identify proteins differentially express between non-infected and ER-infected BAE cells using quantitative proteomics on 1) total proteins extracts and 2) biotinylated membrane proteins extracts. All this experiments were done using high-end MS technology coupled to previous 2D-nano-UPLC separation.

Material and methods

Biological samples

E. ruminantium (Gardel strain, from Guadeloupe, FWI) was produced in bovine aorta endothelial cell (BAE) as previously described (Marcelino *et al.* 2005). Three biological replicates of non-infected and ER-infected BAE cells were used throughout these studies. Total protein extracts were prepared using a SDS-based solubilization protocol and surface protein biotinylation was performed according to Turtoi and co-workers (Turtoi *et al.*). The total amount of protein per biological replicate was quantified using the RC-DC protein assay kit (Biorad) according to the manufacturer's instruction. Biological replicates were then digested overnight using trypsin (ROCHE). The digestion was further extended for 4 h by addition of fresh trypsin.

MS analysis

Trypic peptides obtained from total extracts and biotinylated proteins samples were desalted, desiccated and dissolved in ammonium formiate buffer. Dissolved samples were mixed with MassPREP Digestion Standard Mixture and then injected in the 2D-RPxRP-nano Aquity UPLC (Waters) coupled online with the SYNAPT G2-HDMS qTOF system (Waters). Raw data were processed (deconvolution, deisotoping, protein identification, absolute and relative quantification) using ProteinLynx Global SERVER (PLGS) v2.4 (Waters). Processed MS spectra were searched against *Bos taurus* UniProt database. Ingenuity Systems IPA software™ was then used to analyze and integrate the data obtained from the proteomic assay.

Results and discussion

Total protein extracts

A total of 2,880 proteins isoforms were found to be common trough all replicates for each studied group. Data processing using PLGS allowed identifying the proteins that were up- and down-regulated on total extracts from non-infected and ER-infected BAE cells (Figure 1). The comprehensive analysis of these data was then performed using IPA software in order to perform a global profiling of the identified proteins according to biological networks, canonical pathways and molecular and cellular functions; several protein databases (Uniprot, NextProt, etc) were used to provide more information about the function of the selected proteins. Globally, the analyses performed on the total protein extract indicated that ER induces important changes in host cell namely regarding to the controlled host cell death (apoptosis), by controlling DNA/histone condensation in the host cell nucleus and by modifying cell

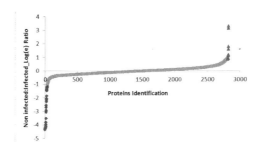

Figure 1. Number of proteins differentially expressed in non-infected and ER-infected BAE cells total protein extracts.

signaling pathways. ER interferes also with the host cell metabolism (particularly for amino acids), host cell osmosis and cytoskeletal organization. These results are associated with the need of the bacteria to expand within its vacuolar compartment in the host cell, inducing the expression of proteins that control the volume of the cell and prevent host-cell lysis due to infection, which is normal because the morula requires a voluminous space in the cell to grow).

Biotinylated extracts

Due to the obvious complexity of the method, several rapid tools for monitoring the inter-replicate repeatability were tested. These consisted on examining the flow-through of the streptavidin purification step for remaining biotinylated proteins, where an internal biotinylated standard was introduced as a control. As supported by the data obtained, the capture of biotinylated proteins was efficient and allowed a good specificity and minor samples loss. Although, the analysis of the biotinylated proteins experiments is still ongoing, preliminary data indicate that a certain number of membranes proteins were immediately observed, as the most intense and with high score, suggesting that BAE labeling with biotin was specific to surface membrane protein.

Conclusions

Until now, no data was currently available on the impact of ER on its host cells. This STSM (Short Term Scientific Mission) revealed for the first time key proteins up-regulated and only presented in infected BAE cells induced by RB replication inside host cell. Although these are preliminary results, requiring additional immunochemical or cellular experiments for full validation, they were crucial to delineate future research projects.

Acknowledgements

The authors acknowledge the financial support from the COST action FA-1002, the Fundação para a Ciência e Tecnologia (FCT, Lisbon, Portugal; contract number PTDC/CVT/114118/2009 and grant # PEst-OE/EQB/LA0004/2011) and the EU project FEDER 2007-2013, FED 1/1.4-30305, 'Risque en santé animale et végétale'. IM acknowledges financial support from the grant SFRH/ BPD/ 45978/ 2008 from FCT. The authors also greatly appreciate the help from Dr. Andrei Turtoi for the preparation of biotinylated extracts.

References

Allsopp, B. A.2010. Natural history of Ehrlichia ruminantium. Vet Parasitol 167: 123-135.

Marcelino, I., de Almeida, A.M., Brito, C., Meyer, D.F., Barreto, M., Sheikboudou, C., Franco, C.F., Martinez, D., Lefrançois, T., Vachiéry, N., Carrondo, M.J.T., Coelho, A.V. and Alves, P.M., 2012. Proteomic analyses of *Ehrlichia ruminantium* highlight differential expression of MAP1-family proteins. Vet Microbiol 156: 305-314.

Marcelino, I., Veríssimo, C., Sousa, M.F.Q., Carrondo, M.J.T. and Alves, P.M., 2005. Characterization of *Ehrlichia ruminantium* replication and release kinetics in endothelial cell cultures. Vet Microbiol 110: 87-96.

Pruneau, L., Emboulé, L., Gely, P., Marcelino, I., Mari, B., Pinarello, V., Sheikboudou, C., Martinez, D., Daigle, F., Lefrançois, T., Meyer, D.F. and Vachiery, N., 2012. Global gene expression profiling of *Ehrlichia ruminantium* at different stages of development. FEMS Immunol Med Microbiol 64: 66-73.

Turtoi, A., Dumont, B., Greffe, Y., Blomme, A., Mazzucchelli, G., Delvenne, P., Mutijima, E.n.N., Lifrange, E., De Pauw, E. and Castronovo, V., 2011. Novel comprehensive approach for accessible biomarker identification and absolute quantification from precious human tissues. J Proteome Res 10: 3160-3182.

Effect of *Staphylococcus aureus* infection on fibrinogen and plasma protein concentrations in experimentally obese rabbits

E. Dishlyanova[1], T.M. Geoggieva[1], T. Vlaykova[2] and I. Penchev Georgiev[1]

[1]Faculty of Veterinary Medicine, Trakia University, 6000 Stara Zagora, Bulgaria; dishlianova@yahoo.com

[2]Faculty of Medicine, Trakia University, 6000 Stara Zagora, Bulgaria

Introduction

The acute phase response to inflammation or infection is encountered in almost all animal species, but the response of individual proteins is species-specific. Kushner *et al.* (Kushner *et al.*, 2006) classify positive acute phase proteins (APP) into 3 groups: (1) APP whose concentration increased by 50% (ceruloplasmin and C_3); (2) APP exhibiting 2- to 3-fold increase (haptoglobin, fibrinogen and α-albumin with antiprotease activity) and 3) proteins, increasing extremely rapidly up to 1000 times (C-reactive protein and SAA). In fact, 98% of protein content of blood plasma (serum) consists of 22 proteins, and the rest 2% comprise about 1000, which could serve as biomarkers (Stanley and Van Eyk, 2005). The changes in plasma proteins concentrations are often utilised for diagnostic purposes (Andonova, 2002; Marshall, 1994).

Staphylococcus aureus is a Gram-positive pathogen able to produce the following toxins: haemolysins, leukocidin, exfoliatin, enterotoxins and TSS-toxin 1 Exfoliating provokes serious skin injuries. The economic losses due to staphylococcal infections in livestock husbandry are a worldwide problem (Devriese *et al.*, 1981). This is particularly valid for rabbi farms, where the infection could be transmitted through newly introduced animals (Devriese *et al.*, 1996).

Acute phase proteins are serum proteins which increase in concentration during the acute phase reaction to infection (Alsemgeest, 1994; Eckersall, 2000). The aim of the present study was to define changes in concentrations of total protein, albumin (as a negative acute phase protein), globuline and fibrinogen (as a positive acute phase protein) in obese and infected with *S. aureus* rabbits. For that, 12 male New Zeland White rabbits, 3 months of age and 2 weeks after castration were divided into 2 groups – experimental and control. 1.5 months after grouning fat, (4 months old rabbits), experimental animals were infected subcutaneously with 100 μl of bacterial suspensionof a field *S. aureus* strain (density: 8×10^8 cfu/ml. The plasma concentration of total protein, albumin, globulins, A/G ratio and fibrinogen were determinated at 3 months, 3.5 months, 4 months = 0 hour before injection and at 6 hours, 24, 48, 72 hours and also days 7, 14, 21 after injection. The protein profiles remained stable in the control group. The concentration of albumin in experimental group decreased significantly after 6 hours to day 7 after infection. The globulin concentrations in the same group significantly increased at 72 hours and day 7 ($P<0.001$) after infection. The albumin/globulins (A/G) ratio in infected and obese rabbits declined significantly after infection compared to the controls after 72 hours to day 7. The fibrinogen concentration in obese, but not infected rabbits (controls) remained stable during all the experimental period. On the other hand, the staphylococcal infection induced gradual and marked increasing in the fibrinogen concentration after 24 hours until day 21. The maximal values – 6.67±0.92 g/l were registered at 48 hours and thereafter this marker slowly declined.

Material and methods

The experiment was performed with 12 male New Zealand White rabbits. At the age of 2.5 months, they were castrated in order to induce obesity. At the age of 4 months, the rabbits were divided into 2 groups of 6 animals each. The experimental group was intradermally infected with 100 µl bacterial suspension of a field *S. aureus* strain (density: 8×10^8 cfu/ml) (Wills *et al.*, 2005). The other group served as control.

Blood samples were collected from *v. auricularis externa* as followed: (1) at the age of 3 months (i.e. 2 weeks post castration); (2) at the age of 3.5 months; (3) at 4 months of age, corresponding to hour 0 before infection. After that, blood samples were obtained on post infection hour 4, 24, 48, 72, and days 7, 14 and 21. The blood plasma was stored at -20 °C.

Total serum protein was assayed using the biuret method (Kolb and Kamishnikov, 1982). Albumin concentration was determined with a commercial kit (Human) containing bromcresol green SU-ALBU INF 156001F, Gesellschaft fur Biochemica, Germany. Globulin concentrations were calculated as difference between total protein and albumin. The albumin to globulin ratio (A/G) was also calculated.

Plasma fibrinogen was determined by nephelometry – method of Podmore with 10% Na_2SO_4 at λ 570 nm (Todorov, 1972).

The data were statistically processed using ANOVA (Statistics for Windows, Stat Soft Ins., USA, 1993). The statistical significance of differences between groups and within a group was evaluated by the post hoc LSD test (Stat Soft Ins., USA, 1993).

Results

Total protein values varied within a narrow range between 3 and 4 months of age, without statistically significant differences within and between the groups ($P>0.05$) The mean levels before infection in obese infected rabbits at 4 months of age were 54.5±2.10 g/l, by the 7th post infection day: 61.9±0.81 g/l, and by the 21st day: 57.4±2.56 g/l.

Albumin concentrations in experimental group decreased considerably between post infection hour 6 and day 7 with lowest concentrations by hour 72 and day 7 ($P<0.001$). Statistically significant differences between groups were detected at 24 hours ($P<0.05$), 48 hours ($P<0.05$), 72 hours ($P<0.01$) and day 7 ($P<0.001$), when albumin levels attained 26.17±1.03 g/l. By post infection days 14 and 21, blood albumin in infected group increased up to levels close to baseline ones.

By 72 hours post infection, globulins in obese infected rabbits increased substantially to 31.1±1.28 g/l ($P<0.001$), and attained 35.71±0.82 g/l ($P<0.001$) by day 7. Globulin concentrations of this group were statistically significantly higher both vs the pre-infection period, as well as vs. controls ($P<0.01$; $P<0.001$).

The A/G ratio was not significantly altered between 3 and 4 months of age. It decreased considerably between 72 hours post infection and day 7 in obese infected animals, with statistically significant difference vs controls and vs baseline pre-infection values ($P<0.001$).

The albumin and globulins percentage in 4-month-old obese rabbits was as followed: albumin: 67.65%; α_1-globulins: 3.21%; α_2-globulins: 5.25%; β_1: 4.10%; β_2: 3.93%; and γ-globulins: 4.63%.

The time course of fibrinogen changes is presented in Figure 1. In both groups, fibrinogen content did not change significantly with age and obesity, as could be seen by the lack of considerable variations between 3 and 4 months of age ($P>0.05$). The *Staphylococcus aureus* had however a significant effect on fibrinogen concentrations in the experimental group, where it increased substantially by 24 hours post infection and significantly higher levels compared to baseline persisted until the 14th day (compared to pre-infection period) and until the 21st day (compared with pre-obesity period). At 3 and 4 months of age the mean fibrinogen values were 2.36±0.21 g/l and 3.06±0.18 g/l ($P>0.05$) respectively, the latter value corresponding to the time of bacterial suspension inoculation. At 24 hours, fibrinogen in experimental rabbits was 5.24±0.48 g/l, by 48 hours – 6.67±0.92 g/l, by 72 hours – 6.05±1.08 g/l, and by the 21st day:

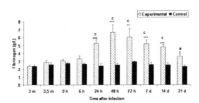

*Figure 1. Concentration of fibrinogen in obese infected (experimental) and obese non infected rabbits (control). Significance of differences between groups: * P<0.05; ** P<0.01; *** P<0.001. Significance of differences in experimental group in comparison the period before obese (3 months): ᵃ P<0.05; ᶜ P<0.001.*

3.62±0.39 g/l. During the entire experimental period, fibrinogen in obese infected rabbits was higher as compared to obese non-infected controls. The differences between both groups were statistically significant by 24 hours post infection ($P<0.001$), 48 ($P<0.001$) and 72 ($P<0.01$), as well as days 7 ($P<0.001$), 14 ($P<0.001$) and 21 ($P<0.05$). The experimentally induced obesity over 1.5 months did not result in any considerable variations in fibrinogen concentrations.

In conclusion, the infection with *S. aureus* of rabbits fattened for a short-time (1.5 months) provided evidence that albumin belonged to slow negative APP, whereas fibrinogen and globulins behaved as positive APP – the fibrinogen was fast-reacting (24 h) and globulins (whose fraction includes all APP) were slow-reacting (48 hours – 7 day).

References

Alsemgeest, S.P.M., 1994. Blood concentration of acute phase proteins in cattle as markers for disease, Utrecht university, the Nederland, Utrecht.

Andonova, M., 2002. Role of innate defence mechanisms in acute phase response against gram-negative agents. Bulg. J. Vet. Med. 5: 77-92.

Devriese, L.A., Godard, C., Okerman, L. and Renault, L., 1981. Characteristics of *Staphylococcus aureus* strains from rabbits. Ann Rech Vet 12: 327-332.

Devriese, L.A., Hendrickx, W., Godard, C., Okerman, L. and Haesebrouck, F., 1996. A new pathogenic *Staphylococcus aureus* type in commercial rabbits. Zentralbl Veterinarmed B 43: 313-315.

Eckersall, P.D., 2000. Acute phase proteins as markers of infection and inflammation; monitoring animal health, animal welfare and food safety. Irish Vet.J 53: 307-311.

Kolb, V.G. and Kamishnikov, V.S., 1982. Practical book in clinical chemistry, 2nd ed, Belarus, pp. 31-33.

Kushner, I., Rzewnicki, D. and Samols, D., 2006. What does minor elevation of C-reactive protein signify? Am J Med 119: e17-28.

Marshall, W., 1994. Plasma proteins. In: W. Marshall (Ed.), An illustrated textbook of clinical chemistry, 2nd ed. Mosby, Mosby-Year Bool Europe Limited, London, UK, pp. 210-221.

Stanley, B. and Van Eyk, J.E., 2005. Clinical proteomics and technologies to define and diagnose heart disease. Walsh. Mech. Cardio 7: 651-665.

Todorov, J., 1972. Nephelimetric determination of fibrinogen (method of Podmore), Clinical Laboratory Technics. Medizina and Fizkultura,, Sofia, pp. 250.

Wills, Q.F., Kerrigan, C. and Soothill, J.S., 2005. Experimental bacteriophage protection against *Staphylococcus aureus* abscesses in a rabbit model. Antimicrob Agents Chemother 49: 1220-1221.

Trypanosoma brucei brucei binds human complement regulatory protein C4BP

Saskia Dolinska[1], Stanislav Hresko[1], Miroslava Vincova[1], Lucia Pulzova[1,2], Elena Bencurova[1], Patrik Mlynarcik[1], Martina Cepkova[1] and Mangesh Bhide[1,2]

[1]Laboratory of Biomedical Microbiology and Immunology, Department of microbiology and immunology, University of Veterinary Medicine and Pharmacy, Kosice, Slovakia; dolinska@uvm.sk

[2]Institute of Neuroimmunology of Slovak Academy of Sciences, Bratislava, Slovakia

Objectives

The uniflagellate protozoan parasite *Trypanosoma brucei* multiplies extracellularly in blood and escapes elimination by the complement system. Activated complement elicits potent biological activities including direct lysis of pathogens (Mollnes *et al.*, 2002; Song *et al.*, 2000; Walport, 2001). One strategy adopted by pathogens to avoid clearance and destruction by complement is to acquire host fluid-phase regulators, like C4b-binding protein or factor H. The acquisition of fluid-phase regulators on the surface of a given pathogen normally results in downregulation of complement activation.

The aim of this study was to examine the interaction of African trypanosomes with human complement regulators.

Material and methods

Trypanosoma *culturing and protein extraction*

The *Trypanosoma brucei brucei* strain (kind gift from Prof. Krister Kristensson, Karolinska Institutet, Sweden) was cultured under standard conditions (37 °C, 5% CO_2 and 90% humidity) in HMI-9 medium supplemented with 20% heat-inactivated fetal bovine serum (Gibco). Cells in log phase from 5th passage were harvested, centrifuged at 2,000×*g* for 15 min at 4 °C, washed twice with ice-cold isotonic buffer (Trizma base 1,4 g, Glucose 1,08 g, NaCl 0.38 g, deionized autoclave water) and resuspended to get a suspension 7.6×10^5 cells/ml. Finally 10 µl of resuspension buffer containing protease inhibitor cocktail (Fermentas) was added. Membrane proteins were extracted using ProteoJET membrane protein extraction kit (Fermentas) according to the manufacturer's instructions. Proteins were separated on 12% SDS-polyacrylamide gels under denaturing conditions and proteins were transferred onto nitrocellulose membrane (Whatman, UK).

Preparation of recombinant complement regulatory proteins

Plasmid pLEXSY_I-blecherry-3 with inserted genes (C1 inhibitor, C4BP, factor H and vitronectin) were electrotransfected into *Leishmania tarentolae*. Recombinant cells were selected with bleomycin at final concentration 100 µg/ml in BHI medium supplemented with hemin and antibiotics (hygromycin, penicillin, streptomycin) in the dark at 26 °C as described before (Hresko *et al.*, 2011). The expression of proteins was induced with tetracycline at the final concentration 10 µg/ml for 48 hrs and expression was checked under microscope at 488 nm. Cells were then harvested by centrifugation for 3 min at 2,000×*g* at room temperature. All proteins were GFP fused.

Affinity ligand binding assay

Membranes were washed once for 5 min in 0.05% TTBS (20 mM TRIS-Cl, pH 7,5; 150 mM NaCl; 0.05% Tween 20, pH 7,4) and blocked for 1 h in blocking buffer (2% skim milk in 0.05% TTBS). Membranes were then incubated for 3 hrs with human serum (1:3 dilution in 1% skim milk in 0.05% TTBS). After two washings with TTBS membrane was incubated for 2 hrs at room temperature in primary antibodies (anti FH in sheep, Abcam; anti human C4BP IgG in sheep and anti human C1 inactivator in sheep, The Binding Site Group; anti vitronectin in mouse, Abcam). All antibodies were diluted 1:150 in 1% skim milk in 0.05% TTBS. Membranes were washed three times for 5 min in wash buffer (0.05% TTBS) and incubated for 1 h at room temperature with secondary antibodies. Secondary antibodies were diluted 1:100,000 in 1% skim milk in 0.05% TTBS. The antibody reactivity was visualized with ECL Western blotting substrate (Pierce) and signals were captured on X-ray film (Kodak Biomax film).

Trypanosoma: CRP interaction by flow cytometry

Trypanosoma brucei brucei was cultured as described above until the cell density reached to 7.6×10^5 cells/ml. One hundred microliters of *Trypanosoma* suspension was incubated in a humidified 37 °C incubator with 5% CO_2 for 1 hrs with 1 µl (~ 100 µg) of recombinant protein (C1 inhibitor, C4BP, FH and vitronectin). After incubation trypanosomes were observed under a microscope to check viability of the cells. Viable parasites were washed two times with ice-cold isotonic buffer and subjected for flow cytometric analysis. *Trypanosoma* cells in HMI-9 medium only and washed with isotonic buffer served as negative control. Event aquisition for flow cytometry was done with following parameters: FL1-H 533/30 nm vs. cell count, flow rate 10 µl/min and 7 µl/analysis. Analyses were performed on DB Accuri C6 flow cytometer (BD Biosciences) and data were processed using FLOWJO software (Tree Star Inc., Ashland, OR).

Results and conclusion

Interaction of C4BP with *Trypanosoma* membrane proteins were confirmed by far-western blotting (Figure 1 Panel A) while interaction with other recombinant proteins (C1 inhibitor, FH,

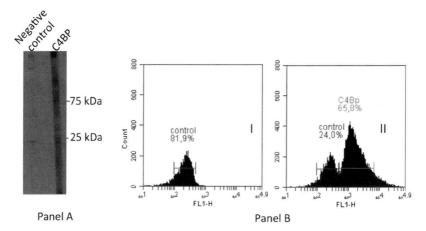

Panel A   Panel B

Figure 1. Interaction between C4b-binding protein and T. brucei brucei. *Panel A – presents binding of C4b-binding protein with ~25 kDa and 75 kDa protein of membrane proteins of* T. brucei brucei. *Negative control – membrane was incubated only with blocking buffer.*
Panel B – I – represents Trypanosoma *cells from the negative control (control gate), II- shows* Trypanosoma *cells incubated with rC4BP-GFP fusion protein. Shift in the FL-1H (C4BP gate) indicates that rC4BP-GFP readily bound on* Trypanosoma.

vitronectin) were not observed with any *Trypanosoma* ligand (data not shown). Trypanosomes incubated in C4BP remained viable. All parasites were motile and of normal shape. Interestingly, the evident binding of rC4BP-GFP fusion protein observed with flow cytometry (Figure 1 Panel B) suggests a possible C4BP mediated complement evasion strategy employed by this parasite. Thus it is tempting to speculate that *Trypanosoma* may use complement-evasion strategies, like other pathogens, which often focus the first line of immune defense: a complement system.

Acknowledgements

This work was supported by competitive academic grants APVV-0036-10 and VEGA–1/0054/12.

References

Hresko, S., Mlynarcik, P., Pulzova, L., Bencurova, E., Mucha, R., Csank, T., Madar, M., Cepkova, M., Bhide, M., 2012. Rapid protein production pipeline in advanced inducible *Leishmania tarentolae* expression system. Farm animal proteomics: 75-79.

Mollnes, T.E., Song, W.C. and Lambris, J.D., 2002. Complement in inflammatory tissue damage and disease. Trends Immunol 23: 61-64.

Song, W.C., Sarrias, M.R. and Lambris, J.D., 2000. Complement and innate immunity. Immunopharmacology 49: 187-198.

Walport, M.J., 2001. Complement. First of two parts. N Engl J Med 344: 1058-1066.

A proteomic analysis of canine serum during the course of babesiosis

Josipa Kuleš[1], Vladimir Mrljak[2], Renata Barić Rafaj[1], Jelena Selanec[2], Richard Burchmore[3] and Peter D. Eckersall[3]

[1]Department of Chemistry and Biochemistry, Faculty of Veterinary Medicine, University of Zagreb, Zagreb, Croatia; jkules@vef.hr
[2]Clinic for internal diseases, Faculty of Veterinary Medicine, University of Zagreb, Zagreb, Croatia
[3]Institute of Infection, Immunity and Inflammation, College of Medical, Veterinary and Life Sciences, University of Glasgow, United Kingdom

Objectives

Canine babesiosis is a tick-borne disease that is caused by the haemoprotozoan parasites of the genus *Babesia* (Taboada and Merchant, 1991). There are limited data of serum proteomics in dogs, and none of babesiosis. The aim of this study was to identify the potential serum biomarkers using proteomic techniques and to increase our understanding about disease pathogenesis.

Materials and methods

Serum samples were collected from 25 dogs of various breeds and sex with naturally occurring babesiosis caused by *B. canis*, admitted to Clinic for Internal Diseases, Faculty of Veterinary Medicine, University of Zagreb, Croatia. Blood was collected at the admission day (day 0), on the first and the 6th day of treatment. Serum was also collected from 10 healthy dogs.

Serum samples were pooled into 4 groups (day 0, day 1, day 6 and control). The protein concentrations of the serum samples were measured using the Bradford assay (Biorad Ltd, Hemel Hempstead UK). Serum from each pool were suspended in rehydration buffer in order to archive final protein concentration of 200 µg/µl. The mixture was loaded onto a 11 cm IPG strip (immobilized pH gradient, pH 3-10, linear, Biorad Ltd, Hemel Hempstead UK). Isoelectric focusing (IEF) was performed (Protean IEF Cell, Biorad Ltd, Hemel Hempstead UK) according to the manufacturers instruction. Electrophoresis in the second dimension was carried out using Criterion precast gels (XT Bis-Tris Gel, 4-12% polyacrylamide gel, IPG+1 well, 11 cm IPG strip; Biorad Ltd, Hemel Hempstead UK) at 200 V for 45-50 minutes. 2-DE gels were then stained with colloidal Coomassie blue G250 and destained in 5% v/v acetic acid. Each sample was analyzed in triplicate. The stained gels were scanned and digitized images of gels were analyzed by using the Progenesis Same Spot software (Nonlinear Dynamics Ltd, Newcastle, UK) to identify protein spots which were differentially expressed through time (power >0.8 and ANOVA significance score of <0.05 between replicate gels). Six selected spots were excised manually and subjected to trypsin digest prior to identification via ion trap mass-spectrometry (Bruker Daltronics Ltd, UK). MS spectra were analyzed with MASCOT software package version 2.2.06 (Matrix Science, UK) to identify spots from an updated sequence database (NCBInr 2012/03/01) using default parameter settings.

Results and discussion

2D electrophoresis of pooled serum samples of dogs with naturally occurring babesiosis (day 0, day 1 and day 6) and healthy dogs were run in triplicate. 2D image analysis showed 64 differentially expressed spots with ANOVA $P \leq 0.05$ and 49 spots with fold change ≥ 2. All together, there were 37 spots with $P \leq 0.05$ and fold change ≥ 2, with power of >0.8. Six spots were selected for running on mass spectrometry (Figure 1).

In this study, we used a proteomic approach to analyze the alterations in canine serum proteome due to *B. canis* infection. 2DE gels on three replicates generated from the infected serum samples were compared with the gels from healthy serum samples, followed by MS/MS analysis, which revealed several significantly differential serum proteins.

A number of differentially expressed serum proteins involved in inflammation mediated acute phase response, complement and coagulation cascades, apolipoproteins and vitamin D metabolism pathway were identified in dogs with babesiosis (Table 1).

Our findings confirmed two dominant pathogenic mechanisms of babesiosis, haemolysis and acute phase response (Reyers *et al.*, 1998; Taboada and Lobetti, 2006). Three APP involved in

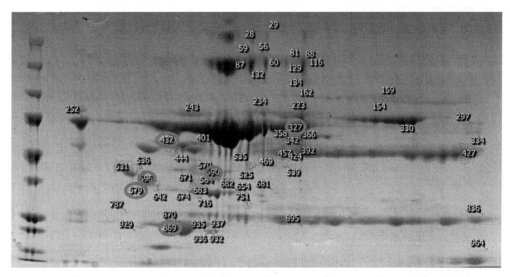

Figure 1. Reference image 2DE map of differentially expressed spots with marked selected spots.

Table 1. List of differentially expressed proteins identified in serum of dogs with B. canis *and healthy dogs with their biological functions.*

Protein name	Biological process	Spot no.
Alpha-1-acid glycoprotein	Acute phase response	596, 579
Apolipoprotein A-I	Cholesterol, lipid and steroid metabolism	869, 432
	Lipid transport	
Antithrombin-III	Blood coagulation	432
Alpha-1-antitrypsin	Acute-phase response	432
Vitamin D-binding protein	Vitamin D metabolic process	432
Apolipoprotein A-IV	Removal of superoxide radicals	590
	Lipid metabolism	
Complement C3	Complement activation (classical and alternative pathway)	590
	Inflammatory response	
Serotransferrin	Cellular iron ion homeostasis	327
	Iron ion transport	
Hemopexin	Cellular iron ion homeostasis	327
Alpha-2-HS-glycoprotein	Acute phase response	596, 432
Haptoglobin	Cellular iron ion homeostasis	579
	Acute phase response	
Alpha-2-antiplasmin	Acute phase response	327
	Fibrinolysis	
Clusterin	Apoptosis complement pathway	596, 579
Leucine-rich-α2-glycoprotein	Acute phase response	432
Inter-alpha-trypsin inhibitor H4	Acute phase response	579

haemoglobin and iron metabolism and transport, haptoglobin, hemopexin and serotransferrin, indicate the role of haemolysis in the course of babesiosis. Tissue hypoxia, which is a common feature in babesiosis, is probably one of the major causes for the release of cytokines, oxygen free radicals, nitric oxide and other inflammatory mediators (Crnogaj *et al.*, 2010; Matijatko *et al.*, 2007). As a consenquence, acute phase response was triggered and demonstrated throughout wide variety of acute phase proteins (alpha-1-acid glycoprotein, alpha-1-antitrypsin, inter-alpha-trypsin inhibitor H4 (ITIH4), leucine-rich-alpha-2-glycoprotein, haptoglobin, hemopexin, serotransferrin, alpha-2-HS-glycoprotein, albumin). Also, release of ROS and consequent oxidative damage, lead to increased expression of clusterin and apolipoproteins (apoA-I and apoA-IV) with antioxidative activity. Impairment of coagulation and fibrinolytic system was demonstrated throughout consumption of AT III due to the coagulation activation. Complement activation was confirmed by clusterin and C3 increased expression. And finally, vitamin D binding protein, as a novel biomarker for MODS in different conditions, can be a possible target for further validation in babesiosis.

These results may provide possible serum biomarker candidates for clinical monitoring and prognosis of babesiosis. The major limitation of this study is insensitivity of 2-DE gel methods used for proteome analysis. Thus, the depletion of major proteins has been suggested to be a potential strategy for enhancing detection sensitivity in serum. However, this study could serve as the basis for further proteomic investigations in canine babesiosis.

Acknowledgements

The present work was carried out at School of Medicine, Veterinary and Life Sciences and Glasgow Polyomics Facility, University of Glasgow, Scotland, UK. Financial support of participating in this project was provided by The British Scholarship Trust. This research was also partially supported by the Ministry of Science, Education and Sports of the Republic of Croatia (Project No. 053 – 0532266 – 2220).

References

Crnogaj, M., Petlevski R., Mrljak V., Kiš I., Torti M., Kučer N., Matijatko V., Saćer I. and Štoković I., 2010. Malondialdehyde levels in serum of dogs infected with *Babesia canis*. Veterinarni Medicina 55, 163-171.

Matijatko, V., Mrljak V., Kiš I., Kučer N., Foršek J., Živičnjak T., Romić Z., Šimec Z. and Ceron J.J., 2007. Evidence of an acute phase response in dogs naturally infected with *Babesia canis*. Veterinary Parasitology 144, 242-250.

Reyers, F., Leisewitz A.L., Lobetti R.G., Milner R.J., Jacobson L.S. and Van Zyl M., 1998. Canine babesiosis in South Africa: more than one disease. Does this serve as a model for falciparum malaria? Annals of Tropical Medicine and Parasitology 92, 503-511.

Taboada, J. and Lobetti R., 2006. Babesiosis. In: Green, C.E. (ed) Infectious Diseases of the Dog and Cat. WB Saunders Co. St Louis, pp. 722-735.

Taboada, J. and Merchant S.R. 1991. Babesiosis of companion animals and man. Veterinary Clinics of North America: Small Animal Practice 21, 103-123.

The acute phase reaction in goats after experimentally induced *E. coli* mastitis: a proteomic approach

Lazarin Lazarov[1], Teodora M. Georgieva[2], Ivan Fasulkov[3], Francesca Dilda[4], A. Scarafoni[5,6], Laura Azzini[5], and Fabrizio Ceciliani[4,6]

[1]Department of Internal Non-infectious Diseases, Faculty of Veterinary Medicine, Trakia University, 6000 Stara Zagora, Bulgaria

[2]Department of Pharmacology, Animal Physiology and Physiological Chemistry, Faculty of Veterinary Medicine, Trakia University, 6000 Stara Zagora, Bulgaria

[3]Department of Obstetrics, Faculty of Veterinary Medicine, Trakia University, 6000 Stara Zagora, Bulgaria

[4]Department of Veterinary Science and Public Health, Università di Milano, 20133 Milano, Italy; fabrizio.ceciliani@unimi.it

[5]Department of Food, Environmental and Nutritional Science, Università di Milano, 20133 Milano, Italy

[6]Interdepartmental Center for Studies on the Mammary Gland – CISMA, Università di Milano, 20133 Milano, Italy

Introduction

Intramammary infections have received a lot of attention because of their major economic impact on the dairy farm. Management strategies, including greater awareness for efficient milking and hygienic measures, have resulted in a significant decrease of clinical mastitis worldwide. Subclinical mastitis, however, continue to be a serious problem.

Escherichia coli causes inflammation of the mammary gland around parturition and during early lactation with striking local and sometimes severe systemic clinical symptoms. Many studies, executed during the last decade, indicate that the severity of *E. coli* mastitis is mainly determined by the host factors rather than by *E. coli* pathogenicity. During *E. coli* mastitis, the host defense status is a cardinal factor determining the outcome of the disease.

Purpose of this study was to determine the diagnostic value of acute phase proteins in goat mastitis. We will focus on two acute phase proteins in particular, Serum Amyloid A (SAA) and α1-acid glycoprotein (AGP). SAA has been widely acknowledged as the major APP in goats (González *et al.*, 2008). AGP, on the contrary, has not been reported to act as an acute phase protein in this species, but it was shown that glycosylation changes can be detected on the surface of AGP purified from inflammatory disease-affected animals when compared to healthy controls (Ceciliani *et al.*, 2009).

The aim of the present investigation was therefore to assess:

1. the possible utilization of SAA and AGP as serum acute phase reactants in goat mastitis; and
2. the presence of post translational modifications on the surface of AGP; as assessed by 2D electrophoresis.

Experimental animals and protocol design

Six mixed bred goats were used in the present study. Prior to the experiment the animals were treated against internal and external parasites. They were fed according to their kind and had free access to tap water. The experimental procedure was approved by the Ethical Committee of Faculty of Veterinary Medicine, Stara Zagora.

The experimental mastitis was induced by intracisternal application of *E. coli.*

Clinical and biochemical analyses

Throughout the experiment, rectal temperature, heart and respiratory rates and changes in the size, shape and consistency of the mammary glands and adjacent lymph nodes, were assessed just before inducing mastitis (hour 0) and 4, 8, 24, 48, 72 and 168 hours after. Blood and milk samples were collected in parallel. Quantitative and qualitative changes in milk and mammary secretions were determined using the California Mastitis Test (CMT). Somatic cell counts (SCC) were also determined daily Blood samples were collected from the puncture of the *v. jugulars* into sterile heparinised tubes, and after centrifugation (1,500 g, 10 minutes, 4 °C) within 30 min after blood collection, plasma was immediately separated and stored at -20 °C until analysis.

Determination of serum APP concentrations

Serum SAA and AGP concentrations were determined by means of a sandwich ELISA test. Cross reactivity of the antibodies was determined by Western Blotting analysis

2D electrophoresis

2-D isoelectric focusing (IEF) sodium dodecyl sulfate-polyacrylamide gel electrophoresis (SDS-PAGE) was performed on 7 cm immobilised pH gradient (IPG) 83 strips (pH 4-7; GE Healthcare) as previously described (Lecchi *et al.*, 2012). Nitrocellulose membranes were the stained with anti-AGP antibodies.

Results and discussion

ELISA analysis on goat serum, were carried out on SAA and AGP. Neither SAA nor AGP increased their serum concentration after *E. coli* challenge.

The second part of the project is still under investigation and was aimed to assess whether post translational modifications of SAA and AGP occur. In particular carbohydrate moiety and different polymerization status will be investigated, for AGP and SAA respectively. Figure 1 presents a Western Blotting of 1 D gel (left) and 2 gel (right) immunostained with specific antibodies against SAA. The figure is representative of the 2D profiles of serum SAA in non-pathological conditions. It is evident from 2D electropherogram that at least two SAA spots are present: the monomer form, with a MW of 12 kDA, and the tetramer, at a MW of 48 kDa. The collection of data on SAA in serum from experimentally infected animals and AGP different glycoforms are still in progress, with the aim to ascertain if post translational occur even if no concentration of the two proteins can be detected in serum.

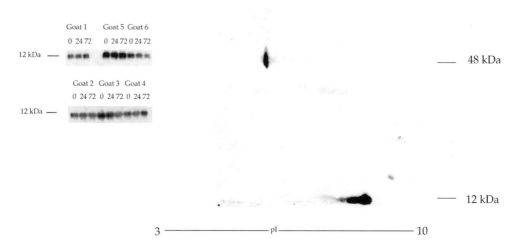

Figure 1. Western Blotting of 1 D gel (left) and 2 gel (right) immunostained with specific polyclonal antibodies raised against SAA.

References

Ceciliani, F., Rahman, M.M., Lecchi, C., Maccalli, M., Pisoni, G. and Sartorelli, P., 2009. Systemic and *in vitro* expression of goat alpha (1)-acid glycoprotein during Caprine Arthritis-Encephalitis Virus infection. Vet Immunol Immunopathol 131: 50-58.

González, F.H., Tecles, F., Martínez-Subiela, S., Tvarijonaviciute, A., Soler, L. and Cerón, J.J., 2008. Acute phase protein response in goats. J Vet Diagn Invest 20: 580-584.

Lecchi, C., Scarafoni, A., Bronzo, V., Martino, P.A., Cavallini, A., Sartorelli, P. and Ceciliani, F., 2012. α(1)-Acid glycoprotein modulates phagocytosis and killing of *Escherichia coli* by bovine polymorphonuclear leucocytes and monocytes. Vet J Aug 18.

The effect of microbial challenge on the intestinal proteome of broiler chickens

Emily L. O'Reilly[1], Richard J. Burchmore[2], P. David Eckersall[1], and Nick H. Sparks[3]

[1]Institute of Infection, Inflammation and Immunity, College of Medicine, Veterinary and Life Sciences, Glasgow University, Glasgow, Scotland, United Kingdom; emilymoo90@hotmail.com
[2]Glasgow Polyomics Facility, College of Medicine, Veterinary and Life Sciences, Glasgow University, Glasgow, Scotland, United Kingdom
[3]Animal & Veterinary Sciences, SRUC Avian Science Research Centre, Auchincruvie, Ayr, Scotland, United Kingdom

Introduction

As well as providing a mechanism by which the body can derive nutrition from its environment, the gastrointestinal tract, composed as it is of physical and chemical immunological components must also safeguard the bird against disease (Yegani and Korver, 2008). Proteomics was used in this study to evaluate the effect of a microbial challenge (in this case used poultry litter) on the intestinal proteome of broiler chickens. Proteomic techniques have only recently been used to study poultry and to date this approach has not been used to study the intestinal proteome in chickens. Difference gel electrophoresis (DIGE), a proteomic approach where the proteins of several samples are labelled with spectrally distinct fluorescent dyes prior to electrophoretic separation (Westermeier, *et al.* 2008) was used to identify proteins differentially expressed in the intestines as a result of exposure to the mixed microbial challenge from used litter. Additionally the pen weights for the groups were recorded and the acute phase protein alpha-1-acid glycoprotein (AGP) a moderate indicator for infection and inflammation was also measured to identify any effects the addition of the litter had on innate immunity and growth rate of the birds.

Materials and methods

At day old, 120 male Ross broiler chicks were divided into groups of 10 and housed in 12 separate pens. On day 12, litter from a unit housing 35 day old broilers was recovered. 1 kg per pen of the used litter was added to 6 pens, randomly assigned as challenge pens. An aliquot of the litter was submitted for microbial analysis using standard methods. On day 12 and days 15, 18 and 22 (3, 6 and 10 days post addition of the used litter) each pen was group weighed and one bird culled and blood and the proximal jejunum recovered. The jejunum was flushed to remove digesta with 0.9% saline and frozen immediately. AGP was measured using a chicken specific single immunodiffusion test kit (ECOS Institute, Miyagi, Japan). Both AGP and pen weights were statistically analysed using Minitab 15. The 6 jejuna from the challenged birds and the 6 from the control birds at each time point were pooled and ground in liquid nitrogen and suspended in DIGE lysis buffer. Acetone precipitation was performed and the protein concentration measured and adjusted to 5mg/ml using a Bradford assay. For each time point (days 12, 15, 18 and 22), 50 μg of intestinal protein from the control and challenged samples were bound separately to both Cy3 and Cy5 fluorescent dyes. A pooled sample containing equal volumes from control and challenged groups at each time point was bound to Cy2 and acted as an internal standard throughout the DIGE experiment. For each time point two gels were ran, with the control and challenged samples bound to opposite Cy dyes, together with the Cy2 internal standard.

For each gel the three Cy bound protein samples were mixed with rehydration buffer containing DTT and IPG buffer and pipetted on to a 24 cm 4-7 IPG strip (GE Healthcare Life Science 17-6002-32). These were focused for 27 hours. For the second dimension the strips were washed in two equilibrium buffers containing DTT and then iodoacetamide for 15 minutes. The strips were then inserted onto a 24 cm SDS gel and ran at 4 V. Gels were scanned on a Typhoon 9400 laser scanner. Images were analysed using Decyder v.7.0 and a two way ANOVA used to select protein spots that were statistically different between the gels. Differentially expressed proteins were excised from preparatory gels loaded with 500ug protein from the same samples, trypsin digested and analysed on Bruker AmaZon ion trap mass spectrometry with comparisons to the MASCOT protein database for protein identification.

Results

Following the microbial challenge from the used litter at day 12, the body weights of the birds challenged tended to be lower than the control birds, the difference becoming significant ($P \leq 0.05$) at day 22, with mean body weights of 374 g (±41 g) and 441 g (±34 g) for the challenged and control birds respectively. There were no statistically significant differences in plasma AGP between birds challenged and not challenged at any time point. A two-way ANOVA analysis was used to access the effects of microbial challenge and sample time point on protein expression. 28 spots were differentially expressed over time as a result of microbial challenge. Of these, 10 showed increased expression in the challenged birds compared to the control birds from day 12, and 16 decreased in the challenged birds from day 12. There were 2 spots that showed variable but significant changes between the two groups over the 4 time points. Figure 1 details the 2D DIGE proteome of the jejunum from the challenged birds at day 22.

Proteins that increased in expression were predominantly actin and actin associated proteins. Proteins in spots 133, 552, 546 and 73 were all found to have both γ-actin and β-actin, both cytoskeletal actins associated with cell motility and cell division. Tropomyosin α-1 chain was also increased and is associated with actin binding and remodelling. The intermediate filament desmin showed a highly significant increase in two spots (102, 552). The calcium binding protein P22 (220), histone binding protein RBBP4 (40) and apolipoprotein A1 (220) also had increased expression in the challenged birds. Of the proteins that decreased in expression in the challenged birds the protein disulfide-isomerase A3 precursor was identified in spots 30, 33, 29 and 77. Three different heat shock proteins showed decreased expression: heat shock

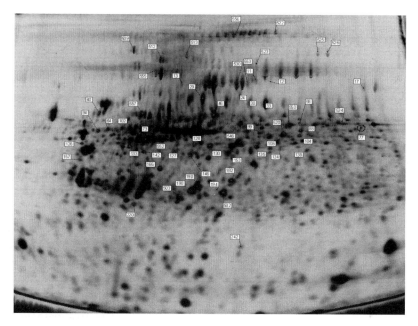

Figure 1. 2D DIGE gel of challenged sample from day 22, 10 days post introduction of litter.

protein 70 and heat hock cognate 71 (530, 531) and heat shock 108 (509). No actin proteins were reduced in expression, though adseverin, a protein that regulates actin filament structure was decreased in the challenged birds. The collagen proteins type IV 1 and 2 were both reduced in expression in the challenged birds.

Conclusions

The introduction of used litter on day 12 had a significant effect on the growth rate of the challenged birds. Though the challenged birds showed no signs of ill health or lethargy, their growth over the 10 day period was adversely effected. Measurement of serum AGP, a moderate and sensitive acute phase respondent in the chicken, indicated no systemic acute phase response was occurring, yet the litter challenge caused a number of significant changes in protein expression over time in the jejunum. Using DIGE to compare the jejunal proteomes of the challenged and control birds in the 10 days following the challenge allowed evaluation of changes in protein expression that are highly localised yet have a significant effect on the growth and possible well being of the birds. Detailing proteomic changes at a localised level may aid in the understanding the factors that contribute to poor growth, enteric disease and reduced welfare and production.

Acknowledgements

This project was funded by the WPSA summer vacation scholarship, and the author would like to thank the WPSA committee, the SRUC and Glasgow Polyomics Facility.

References

Westermeier, R., Naven, T., & Hopker, H. R. 2008 Proteomics in Practice. Wiley-VCH, Weinheim, Germany.

Yegani, M., & Korver, D. R., 2008. Factors affecting intestinal health in poultry. Poultry science, 87, 2052-2063

2D DIGE comparative analysis of *Escherichia coli* strains with induced resistance to enrofloxacin

Cristian Piras[1], Alessio Soggiu[1], Luigi Bonizzi[1], Jarlath Nally[2], Viviana Greco[3], Andrea Urbani[3-4], Piera Anna Martino[1] and Paola Roncada[1,5]
[1]DIVET, Università degli studi di Milano, Milan, Italy; cristian.piras@unimi.it
[2]School of Veterinary Medicine, UCD, Dublin, Ireland
[3]Fondazione Santa Lucia – IRCCS, Rome, Italy
[4]Dipartimento di Medicina Sperimentale e Chirurgia, Università degli Studi di Roma 'Tor Vergata', Italy
[5]Istituto Sperimentale Italiano L. Spallanzani, Milano, Italy

Objectives

Drug resistance in food-borne bacterial pathogens is an almost inevitable consequence of the use of antimicrobial drugs used either therapeutically or to avoid infections in food-producing animals. The considerable use of antibiotics have caused an increased number of antibiotics to which bacteria have developed resistance. Enrofloxacin is one of the most efficient antibiotics against *Escherichia coli* pathogens and there are considerable evidences that document how this microorganism is becoming more resistant to this antibiotic and is developing multidrug resistance (Piras *et al.*, 2012).

Fluoroquinolones are broad-spectrum antimicrobials used in medicine and veterinary practice in Europe for treatment of infectious diseases caused by enteric bacteria such as *E. coli*. Before the early 1990s, fluoroquinolone resistance was rarely found in human and animal *E. coli*; however, since then the frequency has significantly increased worldwide (Hooper, 2001).

Aim of this project is to investigate the proteins involved in enrofloxacin resistance in *E. coli* strains by exposing bacteria to a sublethal concentration. Genomic characteristics of *E. coli* resistant to enrofloxacin have already been documented (Jurado *et al.*, 2008), however poor information have been provided about the proteome of the resistant strains growing in presence of enrofloxacin. The work described represents a comparative proteomic investigation (2D DIGE) of *E. coli* isolates growing under sublethal conditions of enrofloxacin. The main target is to identify through this proteomic approach the proteins involved in this mechanisms of antibiotic resistance.

Material and methods

The experiment was performed on 3 biological replicates for all the 2 experimental groups.

Sample bacterial pellets were washed 4 times in 10 mM TRIS 1mM EDTA to remove the PBS of previous washes. The cell pellet was then weighted and solubilized overnight in a buffer containing 7 M urea 2 M thiourea and 1% ASB14. Protein assay was performed with BioRad Protein assay according to the manufacturer protocol.

For 2DE and 2D DIGE the samples where diluted up to a concentration of 5 µg/µl and adjusted to pH 8.6 adding little amounts of 50 mM NaOH.

2D DIGE experimental design was performed as in following Table 1.

Image analysis was performed with the 2D DIGE module of the Progenesis SameSpots software (Nonlinear Dynamics).

Results and discussion

The experiment was conducted on 3 biological replicates of control and 10 µg/ml resistant *E. coli* isolates.

2D DIGE image analysis revealed the presence of 191 spots with a p value ≤0.05 and 201 with a fold change ≥2. On the Progenesis SameSpots software was applied the filter to save as useful only the spots that had both this characteristics. After the filtering procedure all the spots have been manually checked and 42 have been chosen for the Mass Spectrometry analysis.

The differentially expressed spots that will be analysed through Mass Spectrometry and the spots to cut are shown in Figure 1.

In Figure 1 are also shown some of the most representative proteins differentially expressed among the groups.

Table 1. Design of the DIGE experiment.

Samples	GEL1	GEL2	GEL3
C1	Cy3		
C2		Cy5	
C3			Cy3
E1	Cy5		
E2		Cy3	
E3			Cy5

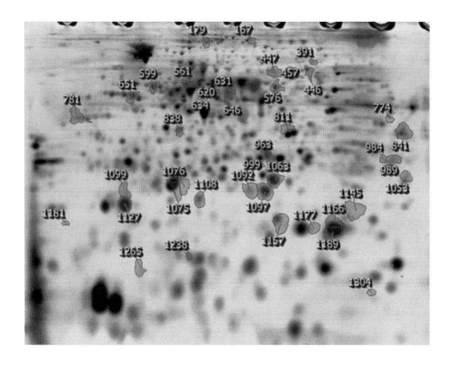

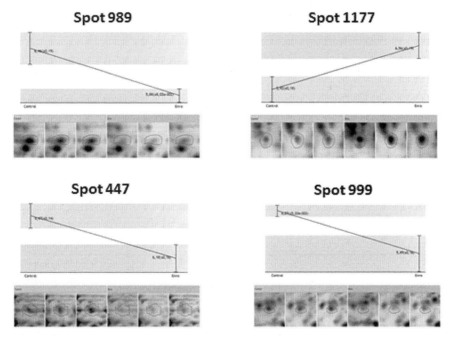

Figure 1. 2D reference map with represented the spots to be analysed through mass spectrometry. All the represented spots had a P-value ≤0.05 and a fold change ≥2. (Spot 989 P-value 0.00119, Spot 1177 P-value 0.00186; Spot 447 P-value 0.00227; Spot 999 P-value 0.00244).

Acknowledgements

Thanks to Prof. Jarlath Nally for the support in performing the 2D DIGE experiments and to the COST Action FA1002 Farm Animal Proteomics for the network provided.

References

Hooper, D.C., 2001. Emerging mechanisms of fluoroquinolone resistance. Emerging infectious diseases 7: 337.

Jurado, S., Orden, J.A., Horcajo, P., De La Fuente, R., Ruiz-Santa-Quiteria, J.A., Martinez-Pulgarin, S. and Dominguez-Bernal, G., 2008. Characterization of fluoroquinolone resistance in *Escherichia coli* strains from ruminants. Journal of Veterinary Diagnostic Investigation 20: 342-345.

Piras, C., Soggiu, A., Bonizzi, L., Gaviraghi, A., Deriu, F., De Martino, L., Iovane, G., Amoresano, A. and Roncada, P., 2012. Comparative proteomics to evaluate multi drug resistance in *Escherichia coli*. Mol Biosyst 8: 1060-1067.

In situ assessment of differential adhesion of neuroinvasive and non-neuroinvasive *Borrelia* strains to BMECs

Lucia Pulzova[1,2], Elena Bencurova[1], Tomas Csank[1], Patrik Mlynarcik[1] and Mangesh R. Bhide[1,2]
[1]Laboratory of Biomedical Microbiology and Immunology, Department of microbiology and immunology, University of Veterinary Medicine and Pharmacy, Kosice, Slovakia; pulzova.lucia@gmail.com
[2]Institute of Neuroimmunology, Slovak Academy of Sciences, Bratislava, Slovakia

Objectives

Invasion of different host niches by *Borrelia* is associated with its translocation across various the body barriers. Neuroinvasive character is in part attributed with crossing of the blood-brain barrier (BBB). This translocation is a multi-stage process with unclear molecular principle. Main prerequisite of successful borrelial traversal seems to be adhesion of *Borrelia* to brain microvascular endothelial cells (BMECs) (Moriarty *et al.*, 2008).

Aim of the study was to assess differential adhesion of neuroinvasive and non-neuroinvasive *Borrelia* strains to BMECs.

Material and methods

Two *Borrelia* strains (neuroinvasive – *Borrelia garinii* strain SKT-7.1 recently classified as *B. bavariensis* and non-neuroinvasive B.b.s.s. serotype 1, strain SKT-2) were grown in BSK-II medium enriched with 6% rabbit serum (Sigma Aldrich) at 33 °C. Antibiotic mixture for *Borrelia* (2 mg phosphomycin, 5 mg rifampicin and 250 µg amphotericin B per ml in 20% DMSO, Sigma-Aldrich) was also added in final concentration 10 µl/ml to avoid the growth of contaminants. Cultures were assessed after one week with dark field microscopy for the presence of *Borrelia* and viability of spirochetes. Well grown cultures (approximately two weeks old) were centrifuged at 2,500×g for 10 min and pellets were resuspended in DMEM medium (Sigma-Aldrich).

Green fluorescent protein and gentamicin resistance were encoded by vector pTM61 (kind gift of Dr. Chaconas and collegues, Canada). pTM61 vector was electroporated into the *E. coli* M15 and transformants were cultured overnight on the selective LB agar containing ampicillin (75 µg/ml), kanamycin (25 µg/ml) and gentamicin (25 µg/ml). Presence of the transformed fluorescent *E. coli* was verified with the fluorescence microscopy. Single *E. coli* colony was inoculated into LB medium, incubated overnight and pTM61 plasmid was purified with QIAprep Spin Miniprep kit (Qiagen).

Vector pTM61 (total DNA 750 ng) was mixed with 50 μl of electrocompetent borreliae and incubated for 1 min on ice. Cell/vector mixture was transferred into cooled electroporation cuvet (0.2 cm electrode gap) and electroporated with single decay pulse of 2.5 kV, 25 μF and 200 Ω with time constant 4-5 ms. Immediately (within 1 min) 1 ml BSK-II medium (room temperature) was added without antibiotics and mixed by pipetting up and down. Cell suspension was transferred into sterile tube that contains 9 ml of BSK-II medium and incubated at 33 °C for 20 hrs. The selection of transformants was achieved with BSK-II medium containing gentamicin (25 μg/ml). Expression of GFP was evaluated by conventional epifluorescence.

Primary cultures of rat BMECs were prepared from 2-week-old Wistar rats, as previously described (Veszelka *et al.*, 2007). Briefly, the forebrains were carefully cleaned of meninges and minced into small pieces and digested with collagenase type II (Worthington Biochemical Corp., NJ, USA) and DNase (Sigma, USA) mixture for 1.5 hrs at 37 °C. The microvessels fragments were separated from myelin layer by gradient centrifugation in 20% bovine serum albumin-DMEM. The microvessels were then digested with collagenase-dispase (Roche Applied Sciences, Switzerland) and DNase (Sigma, USA) for 50 min at 37 °C. Microvessel endothelial cell clusters were separated on a 33% continuous Percoll (Sigma, USA) gradient, collected and washed twice in DMEM. Endothelial cell clusters were then directly plated on collagen type IV (SIGMA) coated (Annexure 1) chamber slide and cultivated in DMEM (Sigma-Aldrich) supplemented with 20% plasma derived serum (First Link, UK), gentamicine, 2 mM L-glutamine, 100 μg/ml heparin (Sigma, USA), 1 ng/ml basic fibroblast growth factor (Roche) and 4 μg/ml puromycin at 37 °C and 5% CO_2 protective atmosphere.

When the BMEC culture was over 90% confluent (assessed with light microscopy) they were challenged with fluorescent neuroinvasive SKT-7.1 and non-neuroinvasive SKT-2 borrelial strains (10^8 cells/ml in 330 μl of DMEM per chamber) resuspended in complete DMEM. Non infected BMECs served as a negative control. Cells were incubated at 37 °C in 5% CO_2 protective atmosphere with constant gentle shaking. After 12 hrs of challenge content of the chambers were removed and non-adherent borreliae were washed out with DPBS three times. Chambers and gasket was detached and slide was fixed with 3% (wt/vol) paraformaldehyde in PBS at room temperature for 30 min. After two times repeated washing with PBS, coverslips were mounted with Fluoroshield (Sigma Aldrich).

Adherent borreliae were visualized using microscope AxioObserver F.1 (Carl Zeiss), software Axiovision (Carl Zeiss) and counted. Assay was performed in triplicate for each strain and Student´s t-test was employed to compare the adherence ability between neuroinvasive and non-neuroinvasive *Borrelia* strains.

Results and discussion

To disclose differential adherence ability of neuroinvasive and non-neuroinvasive *Borrelia*, fluorescent spirochetes were prepared to count the adherent cells simply with conventional epifluorescence microscopy. Both neuroinvasive SKT-7.1 and non-neuroinvasive SKT-2, is able to adhere on the surface of the BMECs after 12 hrs of incubation, and overcome the shear stress, which reflects the situation observed in HUVEC based experiments (Gergel and Furie, 2001). However, significant differences were found in the number of adherent *Borrelia* to BMECs between these two strains *in vitro* (number of adherent SKT-2 were 100 fold less than neuroinvasive SKT-7.1) (Figure 1, Panel A and B).

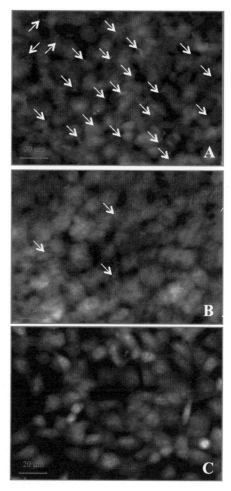

Figure 1. Adhesion to BMEC of neuroinvasive and non-neuroinvasive Borrelia *strain. Cultures of rat BMECs were challenged with neuroinvasive fluorescent strain of SKT-7.1 and non-neuroinvasive strain SKT-2 and incubated under shear stress for 12 hours. Non-adherent borreliae were washed out and adherent borreliae were counted. Adherent borreliae (white arrows) were visualized with fluorescent microscopy. Panel A – BMECs challenged with SKT-7.1. Panel B – BMECs challenged with SKT-2. Panel C – Not challenged BMECs (negative control).*

Results indicate that both strains (SKT-7.1 and SKT-2) possess potential to adhere on BMECs surface; however number of adherent borreliae is dependent on strain. It can be hypothesized, that borrelial ability to adhere on BMECs may influence neuroinvasive character. Difference in level of adherent borreliae may be caused by employment of variety of host adhesive molecules with distinct binding affinity.

Acknowledgements

Work was performed in collaboration with Dr. E. Chakurkar, at ICAR Goa India, mainly for transfection and in-vivo protein production. Financial support was from APVV-0036-10 and VEGA–1/0054/12, 2/0193/11.

References

Gergel, E.I. and Furie, M.B., 2001. Activation of endothelium by *Borrelia burgdorferi in vitro* enhances transmigration of specific subsets of T lymphocytes. Infect Immun 69: 2190-2197.

Moriarty, T.J., Norman, M.U., Colarusso, P., Bankhead, T., Kubes, P. and Chaconas, G., 2008. Real-time high resolution 3D imaging of the lyme disease spirochete adhering to and escaping from the vasculature of a living host. PLoS Pathog 4: e1000090.

Veszelka, S., Pasztoi, M., Farkas, A.E., Krizbai, I., Ngo, T.K., Niwa, M., Abraham, C.S. and Deli, M.A., 2007. Pentosan polysulfate protects brain endothelial cells against bacterial lipopolysaccharide-induced damages. Neurochem Int 50: 219-228.

Caecal IEL, blood lymphocytes and intestinal mucin study in chickens after probiotic prevention and S. Enteritidis infection

Viera Revajova[1], Martin Levkut[1], Maria Levkutova[1], Mikulas Levkut[1,2], R. Herich[1] and Z. Ševčíková[1]
[1]University of Veterinary Medicine and Farmacy, Slovak Academy of Sciences, Bratislava, Slovakia; revajova@uvm.sk
[2]Institute of Neuroimmunology, Slovak Academy of Sciences, Bratislava, Slovakia

Objectives

Salmonella employs multiple, parallel systems of innate immune activation in order to effect a specific series of inflammatory changes. Hosts susceptible to *Salmonella* infection mobilize responses that are necessary for effective control of infection.

In the gut, early host responses to infection include a localized non-specific inflammatory response with an associated influx of heterophils and macrophages which are included in the clearance of bacteria. The timing of clearance in 6-week-old chickens correlates with peak antibody and T cell responses, and considering the luminal location of the *Salmonella*, the specific IgA response is a prime candidate effectors' mechanism.

Probiotics accelerate the establishment of a stable microflora in the gut of neonate chicks and creates the barrier effect against problematic pathogens. The most probiotic provide an antimicrobial effect via lactic acid production and bacteriocins. *Enterococcus faecium* EF55, an isolate from the crop of a chicken (Strompfova *et al.*, 2003), not only possess probiotic character, but it also produces bacteriocin, which probably belongs into II class of them (Franz *et al.*, 2007). EF55 possess genes for enterocins A and P production and has shown an inhibitory activity against target of different bacteria including *Salmonella enterica* (Levkut *et al.*, 2009).

Mucin glycoproteins and secretory IgA are important in the exclusion of pathogens. However, interactions of mucins with antimicrobial molecules and retaining of seretory IgA in mucins are not well understood.

The aim of the study was to evaluate the immunocompetent cells in blood and cecal epithelium (IEL – intraepithelial lymphocytes) together with quantification of jejunal mucin in chickens after *Salmonella* infection. Moreover to follow possible immunostimulatory effect of strain *Enterococcus faecium* EF55 on studied components.

Material and methods

Chickens

Day-old chicks (220) of Cobb-500 hybrids were divided at random into 4 groups (n=55) and placed in large pens with a floor-covering of wood shaving for 11 days. They were fed with commercial diet BR1 and drinking water *ad libitum.*

Experimental design and samplings

The chicks of group 1 (C) were fed only commercial diet without bacteria (negative control). *Enterococcus faecium* EF 55 (Stromfová *et al.*, 2003) in dose 10^9 cfu/ml was perorally administered at 0.2 ml PBS (phosphate buffer saline) to chicks of group 2 (EF) from 0 to 7 days. The birds of group 3 (SE) were inoculated *per os* at day 4 of the experiment with *Salmonella* Enteritidis SE 147 with a single dose 10^8 cfu/ml in 0.2 ml PBS. Chickens of group 4 (EFSE) were administered with *E. faecium* EF 55 during first 7 days in same manner as chicks of group 2, and were inoculated with *S.* Enteritidis at day 4 of the experiment in same way as chicks of group 3. Ten chicks from each group were destroyed on 1, 2, 3, 4 and 7 days p.i. with *Salmonella*. After laparotomy, blood was collected into heparinized tubes by intracardial punction and was immediately used for determination of leukocyte numbers and blood lymphocyte subpopulations. Samples of spleen and caecum for flow cytometry and PCR assays were taken from 5 randomly chosen chicks.

Cecal intraepithelial lymphocyte isolation

Modification of procedure by Swinkels *et al.* (2007) was used. Whole caeca (3 cm) was opened longitudinally, washed 3 times with cold PBS and cut into pieces of 0.50 cm. The suspension of IEL was centrifuged 10 min at 250 g and washed twice in cold PBS. The cells were resuspended in 5×10^4 per 50 μl for imunophenotyping.

Flow cytometry and antibodies

Labelled primary mouse anti-chicken monoclonal antibodies (MoAbs) were used in protocol recommended dilution (Southern Biotechnology Associates, USA). Cells were stained by direct immunofluorescent method at one of the monoclonal combinations CD3-PE/MHC II-FITC, CD4-PE/CD8αα-FITC, TCRγδ-PE/CD45-FITC, IgM-FITC/IgA-PE. Fifty μl of cell suspension and working dilution of specific MoAbs were mixed and incubated 15 min at 4 °C. After staining the cells were measured with FACScan flow cytometer (BD, Germany). The fluorescence data were collected on at least 10,000 lymphocytes and analysed using Cell Quest programme (BD, Germany). The results are expressed as relative percentage of lymphocyte subpopulation, which was positive for a specific MoAb.

Quantification of mucin

Frozen section were stained with PAS. Measurements were taken by light microscope at 400 × magnification. In order to measure the thickness of mucin, the frozen sections were microphotographed (Nikon LABOPHOT 2 with camera adapter DS Camera Control Unit DS-U2, 4x) and then the NIS-Elements version 3.0 softver (Laboratory Imaging, Czech Republic) was used.

Statistical analysis

Analysis of the results was performed using one way ANOVA with Tukey post-test by Minitab 16 (software SC&C Partner Brno, Czech republic).

Results

Number of CD3+ cells in peripheral blood increased in group SE comparing to EF group ($P<0.05$) and controls ($P<0.001$) on 7 day post infection (d.p.i.). Density of CD4+ cells increased in chickens of SE group ($P<0.05$) compare to controls 7 d.p.i. On the contrary, CD8+lymphocytes decreased in the group SE compared to EF and control group ($P<0.001$), and EFSE group ($P<0.01$) on 7 d.p.i. MHCII+ cells decreased in groups treated probiotic ($P<0.05$) compare to control on day 4 p.i. However, on day7 p.i. there was increase in group SE ($P<0.05$) compare to EFSE and control group.

Numbers of intraepithelial MHC+ cells were higher ($P<0.05$) in birds of SE group than in controls and chickens of EF group ($P<0.01$) on 7 day p.i. Intraepithelial IgM+ lymphocytes increased ($P<0.05$) in SE group compared to control chickens on 4 day p.i. There was increase ($P<0.01$) of IgA+ cells in the group EFSE compared to chickens of the group infected with S. Enteritidis on 7 day p.i.

Increased layer of mucin was observed in jejunum of SE group in compare to thickness of mucin in EF group ($P<0.01$), and EFSE group and controls ($P<0.001$) on day 4 p.i. Layer of mucin was thicker in SE group ($P<0.001$) compare to other groups on day 7 p.i. (Table 1).

Table 1. Thickness of jejunal mucin layer (µm; means ± SD).

Days post infection	Control	EF	SE	EFSE
4	12.40±5.58[ae]	15.48±3.68[c]	20.95±2.86[df]	7.74±1.79[be]
7	20.10±9.04[e]	11.61±3.79[ae]	107.70±34.77[f]	9.52±2.39[be]

[ab]$P<0.05$, [cd]$P<0.01$, [ef]$P<0.001$).

Discussion

Salmonella immunization of very young chicks is accompanied by an increase of CD8αα^{+high} γδ T cells in blood, which are activated, and thus represent an important factor for the development of a protective immune response to *Salmonella* organisms. Similarly, the ability of TCR2+ cells migration to the chicken intestine was described, where they provide help to mucosal B cells for IgA production (Davidson *et al.*, 2008). It is known that MHCII molecules can be expressed on the enterocytes especially under inflammatory conditions hence these cells can interact with both CD4+ and CD8+cells (Boismenu and Havran, 1994). Increased thickness of the mucin layer in SE group can be explained not only higher production of mucin but also by admixture of inflammatory exudate and plasma of ruptured edematous enterocytes into mucopolysaccharides.

It can be concluded that preventive administration of *E. faecium* EF55 before *Salmonella* infection showed higher T cell proliferation and T-dependent class switching in B cells from IgM to IgA, that are the beneficial conditions for clearance of *Salmonella*. Moreover, the results of the current study suggest that there is influence of *E. faecium* EF55 on the production of mucin and expression of IgA+ cells with anti-inflammatory effect in birds. Reduction in jejunal mucin layer suggests higher nutrition absorption by birds and beneficial effect of our probiotic on feed conversion. However, it remains an open question about the correlation of increased number of IgA+ cells and optimalisation of mucin layer in biofilm formation.

Acknowledgements

This project was supported, in part, by the Research and Development Support Agency, Slovakia, Grant No. APVV-0302-11, VEGA Slovakia – Grant No. 1/0886/11, No. 1/0313/12, and No.1/0885/11.

References

Boismenu, R. and Havran, W.L., 1994. Modulation of epithelial cell growth by intraepithelial gamma delta T cells. Science 266: 1253-1255.

Davidson, F., Kasper, B., Schat, K:A., 2008. Avian Immunology, First edition. Academic Press, Elsevier, London, pp. 243-249.

Franz, C.M., van Belkum, M.U., Holzapfel, W.H., Abrioure, I.H., Gálvez, A., 2007. Diversity of enterococcal bacteriocins and their grouping in a new classification scheme. FEMS Microbiol Rev 31: 293-310

Levkut, M., Pistl, J., Lauková, A., Revajová, V., Herich, R., Ševčíková, Z., Stromfová, V., Szabová, R., Kokinčáková, T. 2009. Antimicrobial activity of *Enterococcus faecium* EF55 against *Salmonella* Enteritidis in chickens. Acta Vet Hung 57: 13-24.

Strompfová, V., Mudroňová, D., Lauková, A. 2003. Effect of bacteriocin-like substance produced by *Enterococcus faecium* EF55 on the composition of avian gastrointestinal microflora. Acta Vet Brno 72: 559-564.

Swinkels, W.J., Post, J., Cornellisen, J.B., Engel, B., Boersma, W.J., Rebel, J.M. 2007. Immune response to an *Eimeria acervulina* infection in different broilers lines. Vet Immunol Immunopathol 117: 26-34.

Bovine anaplasmosis: IgG2 reactivity in vaccinated cattle against conserved recombinant outer membrane proteins

Theah Molad, Ludmila Fleiderovitz and Varda Shkap
Division of Parasitology, Kimron Veterinary Institute, P.O. Box 12, Bet Dagan 50250, Israel;
moladt@int.gov.il

Objectives

Anaplasmosis is a vector- borne infectious blood disease of cattle caused by the intra-erythrocytic bacterial pathogen *Anaplasma marginale*. Currently, the only effective method of immunization is achieved by immunization of cattle with live vaccine *Anaplasma centrale* strain, which causes mild form of anaplasmosis (Molad *et al.*, 2006; Shkap *et al.*, 2008). However, the combined data of complete genome sequences and proteomic analyses of *A. marginale* and *A. centrale* nowdays provide information that enables the identification of *Anaplasma* antigens potentially related to protective immunity. Aim of the study was to express *A. marginale* and *A. centrale* recombinant subdominant conserved proteins, and to assess the antibody reactivity in sera of *A. centrale* vaccinated cattle against the expressed proteins.

Materials and methods

The genes encoding for the *A. marginale* AM779, VirB9-1, VirB9-2, VirB10 and elongation factor-Tu (EF-Tu) proteins of the Israeli most prevalent field strain, and the gene encoding for the vaccine *A. centrale* outer membrane protein 7 (Omp7), were amplified using specific primers. Following the PCR, the amplified products were cloned in frame into expression vector pET-28b (Novagen CA, USA). The pET-28b constructs were transformed into competent *Escherichia coli* BLR (DE3), BL21 or Rosetta cells (Novagen, CA, USA). Cultures were grown to an optical density of 0.5-0.7 (OD_{600}) in Luria-Bertani medium containing 50 μg of kanamycin per ml, and then induced with 1 mM isopropyl-ß-D-thiogalactopyranoside (IPTG). The recombinant proteins obtained, rAM779, rVirB9-1, rVirB9-2, rVirB10 and rEF-Tu were solubilized under denaturing conditions with 6 M guanidine hydrochloride. Ni-NTA agarose beads were used for purification of the His-tagged proteins. The eluted proteins were dialyzed at 4 °C against refolding buffer containing 6, 4; 2 M urea and finally no urea was applied. During dialysis the *A. centrale* Omp7 (ACOmp7) was precipitated, therefore the protein was solubilized first 6 M guanidine hydrochloride, then mixed with Ni-NTA beads and further washed with buffer containing 100 mM NaH_2PO_4, 10 mM TrisHCl, 8 M urea, pH 8.0. After washings, 20× basic buffer containing 25 mM TrisHCl, 0.5 M NaCl and 10%, glycerol, pH 8.0 was added to the resin-protein mixture. The protein was allowed to refold at 4 °C overnight, and at the next day washings were performed with the buffer containing 50 mM NaH_2PO_4, 300 mM NaCl, 20 mM imidazole at pH 8.0. The protein ACOmp7was finally eluted with elution buffer (250 mM NaH$_2$PO$_4$, 300 mM NaCl, 20 mM imidazole, pH 8.0) and 0.5 M arginine.

MaxiSorp plates (Nunc, Roskilde, Denmark) were pre-coated with the purified recombinant proteins at a concentration of 1 µg/well in a coating buffer containing 50 mM carbonate buffer, pH 9.6, at 4 °C overnight to perform indirect-ELISA: The plates were washed with phosphate-buffered saline (PBS) -0.05% Tween (PBS-T), pH 7.4, and incubated with blocking solution (10% donor horse serum in PBS-T) for 1 hour at 37 °C. After washing test serum samples diluted from 1:100 to 1:2,000 in blocking solution were added, for 2 hours incubation at 37 °C. Following washings as above, sheep anti-bovine IgG2 horseradish peroxidase conjugates (Serotec, Dusseldorf, Germany) diluted 1:10,000 in blocking buffer in PBS-T were added. The plates were incubated for 1 hour at 37 °C and then washed with PBS-T. The substrate BM Blue POD 3, 3´-5, 5´-tetramethylbenzidine (TMB) (Roche, Mannheim, Germany) was added for another 30 min of incubation. The reaction was stopped by the addition of 1 M sulfuric acid. The optical density was measured at 450 nm.

The recombinant proteins were tested for reactivity with serum samples from *A. centrale* vaccinated calves collected up to 20 weeks after vaccination. In two calves 405 and 429 the follow up for the antibody reactivity was performed from week 8 to after vaccination. In parallel, sera from *A. marginale* infected calves were examined.

Results and discussion

In the present study, surface exposed proteins VirB9-1, VirB9-2, VirB10, of the type IV secretion system, EF-Tu, AM779 and ACOmp7 were expressed, IgG2 antibody titers specific to the recombinant protein constructs were detected in *A. centrale*-immunized calves. The reciprocal antibody titers to *A. marginale* and *A. centrale* antigens ranged from 1:100 to 1:1000. In calf 405 (Table 1) the IgG2 responses were detected from week 8 after immunization up to week 50, while an increase in the antibody response against *A. marginale* recombinant proteins rVirB9-1, rVirB9-2, rVirB10, rEF-Tu and rAM779 was observed from week 39 post vaccination. IgG2 titers specific to recombinant *A. centrale*, rAComp7 antigen showed higher reactivity in most of the *A. centrale* vaccinates, as compared to the titers against the other proteins tested (Table 1). Similarly, seroreactivity against the recombinant proteins was detected in *A. marginale* infected calves, with antibody titers ranging from 1:100 to 1:1000.

The IgG2 antibody response in cattle is known to enhance phagocytosis and complement fixation cascade reactions, and thus augment the protective immunity against *Anaplasma* infection (Brown *et al.*, 1998). Using genomic and proteomic approaches minor components of the outer membrane protein were identified (Palmer *et al.*, 2011). The expressed proteins AM779, VirB9-1, VirB9-2, VirB10, EF-Tu and Omp7 reactive with *A. marginale* immune sera were originally identified by mass spectrometry followed by elution of proteins from two-dimensional gel spots (Lopez *et al.*, 2005). Although markedly less abundant, these

Table 1. Detection of IgG2 in sera samples from A. centrale *- vaccinated calf 405 by an indirect ELISA.*

ACOmp7	EF-Tu	VirB10	VirB9-2	VirB9-1	Am779	Weeks post infection
8	100	100	200	200	100	neg
21	100	200	200	200	100	500
26	100	200	200	200	100	400
27	100	200	200	200	200	400
35	200	200	200	200	200	500
39	400	400	400	500	300	500
43	400	500	500	500	300	500
46	500	500	500	500	300	1000
50	500	500	500	500	400	1000

minor proteins do not vary during infection and are highly conserved among *A. marginale* strains. The six proteins expressed in this study were confirmed as being immunologically 'subdominant' antigens, and are conserved between *A. marginale* and the vaccine *A. centrale* strain. Recognition of the recombinant proteins in the present study by IgG2 antibodies from *A. centrale-* vaccinated cattle indicates that these proteins might be useful protective antigens candidates for the development of recombinant vaccine against *A. marginale*.

Acknowlegments

This work was supported by the USA-BARD 4187-09C grant.

References

Brown, W.C., Shkap, V., Zhu, D., McGuire, T.C., Wenbin Tuo, McElwain, T. F. and Palmer, G.H., 1998. CD4(+) T-lymphocyte and immunoglobulin G2 responses in calves immunized with *Anaplasma marginale* outer membranes and protected against homologous challenge. Infect Immun 66: 5406-5413.

Lopez, J. E., Siems, W. F., Palmer, G.H., Brayton, K. A., McGuire, T. C., Norimine, J. and Brown, W.C., 2005. Identification of novel antigenic proteins in a complex *Anaplasma marginale* outer membrane immunogen by mass spectrometry and genomic mapping. Infect Immun 73: 8109-8118.

Molad, T., Mazuz, M.L., Fleiderovitz, L., Fish, L., Savitsky, I., Krigel, Y., Leibovitz, B., Molloy, J., Jongejan, F. and Shkap, V., 2006. Molecular and serological detection of *A. centrale*- and *A. marginale*-infected cattle grazing within an endemic area. Vet Microbiol 113: 55-62.

Palmer, G. H., Brown, W.C., Noh, S. M. and Brayton, K. A., 2012. Genome-wide screening and identification of antigens for rickettsial vaccine development. FEMS Immunol Med Microbiol 64: 115-119.

Shkap, V., Leibovitz, B., Krigel, Y., Molad, T., Fish, L., Mazuz, M., Fleiderovitz, L. and Savitsky, I., 2008. Concomitant infection of cattle with the vaccine strain *Anaplasma marginale ss centrale* and field strains of *A. marginale*. Vet Microbiol 130: 277-284.

Identification BMECs receptors interacting with *Trypanosoma brucei brucei*

Miroslava Vincova[1], Andrej Kovac[2], Lucia Pulzova[1,2], Saskia Dolinska[1] and Mangesh R. Bhide[1,2]
[1]*Laboratory of Biomedical Microbiology and Immunology, Department of microbiology and immunology, University of Veterinary Medicine and Pharmacy, Kosice, Slovakia;*
vincova.miroslava@gmail.com
[2]*Institute of Neuroimmunology, Slovak Academy of Sciences, Bratislava, Slovakia*

Objectives

Neural Trypanosomiasis is fatal disease caused by *Trypanosoma brucei*. *T. brucei* complex consists of three major subspecies namely *T. brucei brucei*, *T. brucei gambiense*, *T. brucei rhodesiense*. They are transmitted by the saliva of the infected tsetse flies of genus *Glossina* during blood meal. Trypanosomas are extracellular, spread through the blood stream to various niches, and consequently invade the central nervous system (Masocha *et al.*, 2007). It is necessary to know how *T. brucei* interacts with brain microvascular endothelial cells (BMECs) to cross the blood brain barrier. Previous studies have shown that majority of extracellular pathogens have to adhere to BMEC first (transient adhesion), which then induces series of cell events causing more firm adhesion of pathogen to BMECs (stationary adhesion), followed by alteration in the extracellular matrix near tight junctions allowing more space for pathogen to cross BBB (reviewed in Bencurova *et al.* 2011). To this background, the study was aimed to identify putative surface adhesion receptors on BMECs that may take part in the *Trypanosoma*:BMEC interface during transient adhesion. Unfolding the underlying principles of ligand:receptor interactions, which take place during passage of *Trypanosoma* through BBB is important to understand its neuroinvasive mechanisms.

Material and methods

Trypanosoma *culturing*

The *T. brucei brucei* (kind gift from Prof. Krister Kristensson, Karolinska Institutet, Sweden) was cultured in HMI-9 medium (Hirumi's modified Iscove's medium) at 37 °C under aerobic conditions (5% CO_2). Shortly, *Trypanosoma* low passage conserve was serially diluted 1:4 in HMI-9 in 24 well plate and incubated for 3 days. Cultures were monitored once a day for viability and growth. 10-15 wells having density of *Trypanosoma* cells $\sim 2\text{-}4\times10^5$ were selected, serially diluted 1:4 in fresh HMI-9 medium and incubated for next 3 days. Content of each well (500 µl) was then transferred to twenty 25 cm^2 flasks containing 3.6 ml of fresh HMI-9 each and incubated for next 2 days.

Cultivation of BMECs and extraction of surface proteins

BMECs were prepared from Wistar rats as described previously (Pulzova *et al.*, 2009; Veszelka *et al.*, 2007). Forebrains were cleaned of meninges, minced into small pieces and digested with collagenase type II and DNase. Fragments were separated from myelin layer by centrifugation and microvessels were digested with collagenase-dispase and DNase. Microvessel endothelial cell clusters were separated on a 33% continuous Percoll and endothelial cell clusters were then directly plated on collagen type IV coated chamber slide. Cells were cultivated in DMEM supplemented with 20% human plasma derived, gentamicine, L-glutamine, heparin, 1 basic fibroblast growth factor and puromycin. Cells were harvested and surface proteins were extracted with Proteojet membrane protein extraction kit as per manufacturer's instructions (Thermo scientific, Slovakia).

Hybridization of BMEC proteins with Trypanosoma cells and surface shaving

When *Trypanosoma* culture achieved density ~ $4{\times}10^5$ (early exponential growth phase), 15 ml of culture (content of four 25 cm^2 flask) was centrifuged at 2,000×g for 15 min at 4 °C. Approximately 13 ml was removed and pellet was resuspended in remaining medium. 100 µl of trypsin solution containing different trypsin concentrations (0.05%, 0.1%, 0.25% and 0.5%, Sigma-Aldrich) was added to standardize optimal concentration for shaving.

To hybridize BMEC proteins with *Trypanosoma*, surface proteins (750 µg in 100 µl of ultrapure water) extracted from BMECs were added to the culture and incubated for 1 hr at 37 °C at 5% CO_2. For negative control, 100 µl of ultrapure water was added and cells were incubated as cited earlier. *Trypanosoma* cells were then washed two times with isotonic buffer (Trizma base 1.4 g, Glucose 1.08 g, NaCl 0.38 g), finally pellet was resuspended in 1 ml isotonic buffer containing 0.25% trypsin and incubated at 37 °C for 30 min. After incubation cells were pelleted and supernatant was transferred to new tube. Peptides were precipitated by adding 3 volumes of ultrapure methanol or acetonitrile and subjected for mass spectrometry. *Trypanosoma* pellet was simultaneously checked under dark field microscopy to confirm intactness of the cells. Whole assay was repeated at lest three times.

Protein identification in shaving

Supernatants were evaporated to dryness under vacuum and residues were subjected to nano-UHPLC (Dionex). The mobile phase consisted of solvent A (2% acetonitrile in water with 0.05% TFA) and solvent B (80% acetonitrile in water with 0.05% TFA). Eluted fractions were spotted on MALDI targets (96 spots per run). MS and MS/MS spectra were collected by ultrafleXtreme (Bruker) and the proteins were identified by MASCOT software (Matrix Science).

Results and conclusion

Among different trypsin concentrations 0.25% was the best, which did not lysed *Trypanosoma* cells. Majority of the cells were motile under dark-field microscopy observed after shaving.

When peptides after tryptic (0.25% trypsin) shaving of *Trypanosoma* cells were precipitated with methanol total 77 protein candidates were identified with MALDI mass spectrometry against 39 when acetonitrile was used. Thus methanol was used further for shaving of BMEC proteins bound to *Trypanosoma* cells.

Protein shavings followed by mass spectrometry performed to identify BMEC surface proteins bound to *Trypanosoma* cells revealed an array of protein candidates. After evaluating each protein from 97 identified candidates, redundant proteins were discriminated based on their subcellular localization and known function; and the most probable protein candidates were shortlisted (Table 1). All five proteins possess potential of being expressed on outer cell membrane of BMECs. Among these proteins, member of interferon-induced transmembrane protein family modulate cell adhesion and influence cell differentiation (Tanaka *et al.*, 2005). Interferon-α and -β induce the expression of IFITM3 while interferon-γ up-regulates IFITM1 (Gutterman, 1994). It is tempting to speculate the role of moesin in the adhesion or interaction with *Trypanosoma* as ezrin, radixin and moesin (ERM) are found to regulate cortical morphogenesis and cell adhesion by connecting membrane adhesion receptors to the actin-based cytoskeleton (Barreiro *et al.*, 2002). Annexin, on the other hand, takes part in interactions with various cell-membrane components that are involved in the structural organization of the cell and intracellular signaling by enzyme modulation and ion fluxes (Moss and Morgan, 2004).

Furhter experiments to confirm interaction between *Trypanosoma* surface ligands and five adhesion receptors of BMEC found in this study is on going. To our knowledge there is no previous report available that shows such interaction. Confirmation of ligand:receptors

Table 1. The most probable protein candidates.

Accession	Protein	Scores	MW [kDa]
IFM3_RAT	Interferon-induced transmembrane protein 3 OS=*Rattus norvegicus* GN=ifitm3 PE=2 SV=1	101.5 (M:101.5)	15.0
ANXA1_RAT	Annexin A1 OS=*Rattus norvegicus* GN=Anxa1 PE=1 SV=2	71.8 (M:71.8)	38.8
ANXA2_RAT	Annexin A2 OS=*Rattus norvegicus* GN=Anxa2 PE=1 SV=2	50.7 (M:50.7)	38.7
ANXA5_RAT	Annexin A5 OS=*Rattus norvegicus* GN=Anxa5 PE=1 SV=3	51.5 (M:51.5)	35.7
MOES_RAT	Moesin – *Rattus norvegicus*	36.2 (M:36.2)	67.7

interaction will provide deep insight in to the complex interface formation between *Trypanosoma* and BMECs during transient adhesion and BBB translocation.

Acknowledgements

Financial support was from APVV-0036-10 and VEGA−1/0054/12, 2/0193/11. MV perusing her doctoral study under the framework of European structural fund ITMS 26110230036.

References

Barreiro, O., Yanez-Mo, M., Serrador, J.M., Montoya, M.C., Vicente-Manzanares, M., Tejedor, R., Furthmayr, H. and Sanchez-Madrid, F., 2002. Dynamic interaction of VCAM-1 and ICAM-1 with moesin and ezrin in a novel endothelial docking structure for adherent leukocytes. J Cell Biol 157: 1233-1245.

Bencurova, E., Mlynarcik, P. and Bhide, M., An insight into the ligand-receptor interactions involved in the translocation of pathogens across blood-brain barrier, 2011. FEMS Immunol Med Microbiol 63: 297-318.

Gutterman, J.U., 1994. Cytokine therapeutics: lessons from interferon alpha. Proc Natl Acad Sci U S A 91: 1198-1205.

Masocha, W., Rottenberg, M.E. and Kristensson, K., 2007. Migration of African trypanosomes across the blood-brain barrier. Physiol Behav 92: 110-114.

Moss, S.E. and Morgan, R.O., 2004. The annexins. Genome Biol 5: 219.

Pulzova, L., Bhide, M.R. and Andrej, K., 2009. Pathogen translocation across the blood-brain barrier. FEMS Immunol Med Microbiol 57: 203-213.

Tanaka, S.S., Yamaguchi, Y.L., Tsoi, B., Lickert, H. and Tam, P.P., 2005. IFITM/Mil/fragilis family proteins IFITM1 and IFITM3 play distinct roles in mouse primordial germ cell homing and repulsion. Dev Cell 9: 745-756.

Veszelka, S., Pasztoi, M., Farkas, A.E., Krizbai, I., Ngo, T.K., Niwa, M., Abraham, C.S. and Deli, M.A., 2007. Pentosan polysulfate protects brain endothelial cells against bacterial lipopolysaccharide-induced damages. Neurochem Int 50: 219-228.

Effect of short-time overweight gain on the plasma PON1 activity after acute *Staphylococcus aureus* infection in rabbits

Tatyana I. Vlaykova[1], Evgeniya Dishlyanova[2], Ivan P. Georgiev[2] and Teodora M. Georgieva[2]
[1]*Dept. Chemistry and Biochemistry, Medical Faculty, Trakia University, 11 Armeisks Str. Stara Zagora, 6000, Bulgaria; tvlaykov@mf.uni-sz.bg*
[2]*Dept. Pharmacology, Physiology and Physiological Chemistry, Faculty of Veterinary Medicine, Trakia University, Student campus, Stara Zagora, 6000, Bulgaria*

Objectives

The plasma enzyme paraoxonase 1 (PON1) is suggested to be an important antioxidant enzyme that is able to inhibit the oxidation of LDL complexes. PON1 is a HDL-associated enzyme, belonging to the family of calcium-dependent hydrolases (esterases) that catalyzes the hydrolysis of many xenobiotics (Costa *et al.*, 2005; van Himbergen *et al.*, 2006). The antioxidant activity of the enzyme is considered to be executed by hydrolysis of a variety of aromatic and aliphatic lactones, including lactones of hydroxyl derivatives of arachidonic and docosahexaenoic acid and of some oxidative products of the phospholipids and polyunsaturated fatty acids (PUFAs) (Costa *et al.*, 2005; Draganov and La Du, 2004; Mackness *et al.*, 2002). Moreover, there is a strong evidence that PON1 hydrolyzes also PAF (platelet-activating factor), one of the important pro-inflammatory mediators with phospholipid structure. It has been found that the all PAF-hydrolyzing activity of HDL is due to PON1 (Mackness *et al.*, 2002).

Previously we reported that plasma PON1 could be considered as a negative acute-phase protein (APP) in rabbits after infection with *Staphylococcus aureus* (Vlaykova *et al.*, 2013). Bearing in mind that obesity is associated with a state of chronic inflammation (Lumeng and Saltiel, 2011) and considering that in human there is positive correlation of body fat mass with plasma APP (Aydin *et al.*, 2012; Forouhi *et al.*, 2001; Perry *et al.*, 2008), we aimed to explore the effect of short-time obesity on the plasma level of PON1 after infection with *S. aureus*. As a model for obesity we chose a short time overweight gain after castration of male rabbits (Georgiev *et al.*, 2011).

Materials and methods

The experimental procedure was approved by the Ethic Committee at the Faculty of Veterinary Medicine. In the current study we performed experiment with 4 groups of male New Zealand white rabbits: Iex (non-obese infected, n=7), Icon (non-obese controls, n=6), IIex (obese infected, n=6) and IIcon (obese controls, n=6).

The rabbits from infected group were inoculated subcutaneously with 100 µl of bacterial suspension of a high virulent field *S. aureus* strain (density: 8×10^8 cfu/ml) as described by Murray *et al.* (1999). The rabbits from control group were inoculated with saline. Blood samples with heparin were taken from *v. Auricularis*.

The non-obese rabbits from experimental (Iex) and control (Icon) groups were enrolled at age of 3 months and were with average body weight of 3.2 kg. The rabbits from the obese groups (IIcon and IIex) were castrated at age 2.5 months and fed for growing fat for 1.5 month. After that period the mean average body weight of the rabbits at age of 4 months was 4.43 ± 0.23 kg; 6 of those rabbits were inoculated with *S. aureus* strain (IIex) and the rest 6 animals served as controls (IIcon).

The plasma paraoxonase activity (PON activity) of PON1 toward the substrate paraoxon (paraoxonase activity) was evaluated using an adapted by us method of Tomas *et al.* (Emin *et al.*, 2009; Tomas *et al.*, 2000). The method is kinetic and it is based on the velocity of hydrolysis of the paraoxon by the PON1 enzyme. The concentration of PON1 in plasma was presented as U/l. The paraoxonase activity (PON activity) of 1 U was defined as 1 µmol *p*-nitrophenol formed per minute. The molar extinction coefficient of *p*-nitrophenol is 18,053 $(mol/l)^{-1}/cm$ at pH 8.5

Statistical analyses were performed using SPSS 16.0.0 (SPSS Inc.), applying the ANOVA test, Student *t*-test and paired *t*-test. Factors with $P<0.05$ were considered statistically significant.

Results and discussion

The baseline plasma PON1 activities were commensurable in all groups of animals (Iex:1,680±227 U/l; Icon:1,642±107 U/l; IIex-1,824±226 U/l; IIcon:1,675±263 U/l, $P=0.584$) (Figure 1). The level of the enzyme activity in plasma of non-obese infected rabbits decreased significantly at 24 hours reaching only 61% of the baseline level (1,115±115 U/l, $P=0.034$), gradually increase at 48 hours being 78% of the baseline level (1,302±129 U/l, $P=0.061$) and returned to the baseline level at day 7 (1,647±99 U/l, $P=0.956$) and remained higher, although not significantly, than before infection to the end of the follow-up period.

Similar kinetics was seen in the obese-infected group (IIex), however the marginally significant decrease was detected at 72 hours reaching 77% of the baseline level (1,409±294 U/l, $P=0.057$), which was followed by increase to the baseline level at day 7 and continued to increase slightly till the end of the follow-up period ($P>0.05$) (Figure 1). No significant changes were observed in the plasma levels of PON1 activity in both control groups of untreated rabbits (Icon and IIcon) during the whole follow-up periods (Figure 1).

Comparing the PON1 activity in different experimental groups at particular studied time periods we found that the concentrations of PON1 of non-obese infected rabbits were

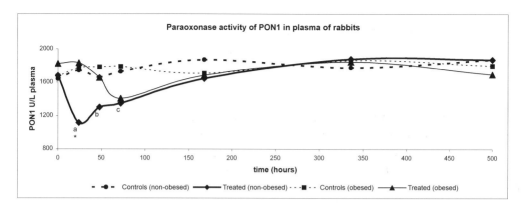

Figure 1. Kinetics of plasma paraoxonase activity of non-obese infected, obese-infected, non-obese control and obese control rabbits. Data are presented as mean ± standard deviation (mean ± SD). (a: P<0.001, between Iex group and all other animal groups; b: P<0.05, between Iex group and all other animal groups; c: P=0.022, between IIex group and IIcon; *: P=0.034, Iex at the 24th hour vs. baseline level).

significantly lower at 24 hours (P<0.001) and at 48 hours (P<0.05) and remain to be lower although not significantly at 72 hours than the enzyme levels of all other three groups (Icon, IIcon and IIex). Concerning the PON1 plasma concentration of obese infected (IIex) rabbits, we found that the enzyme level was significantly lower than that of the obese controls only at 72 hours (P=0.022). At that period PON1 activity of IIex group was also lower but not significantly comparing to the obese non-infected controls (IIcon) (P=0.072). There was no significant difference in the plasma PON1 activity at any of the time periods between the animals of the two control groups (non-obese, Icon and obese, IIcon).

One of the findings in our study is that the short-time overweight gain in male rabbits after castration does not change the level of paraoxonase 1 activity in plasma, which could propose no effect of experimentally induced obesity on the liver production of that enzyme. In addition, the results of our study prove that acute *S. aureus* infection in rabbits causes a decrease in plasma paraoxonase activity both in animals with normal weight and with short-time overweight gain. These latter findings confirm that PON1 could be considered as negative acute-phase protein in rabbits after acute Gram-positive bacterial infection. However, the presence of overweight (experimental obesity) led to relatively later and less grade of reduction of plasma PON1 activity. Thus, we can suggest that the short-time overweight gain after castration might suppress the acute phase response to bacterial infection. Our observations are in line with the previously reported effect of the same type of short-time overweight gain on ceruloplasmin as positive APP (Georgieva *et al.*, 2012). However, in an obese model mice, the high-fat diet-induced obesity was reported to result in higher levels of the LPS binding protein (LBP), serum amyloid A (SAA) and RANTES one day after LPS infusion compared with normal-fat fed control mice (Nakarai *et al.*, 2012). This and other reports suggest that obesity is associated with heightened acute inflammatory response. The difference in the findings with those of our study might me attributed to the type of infection agents and animal species. We propose that the decreased level of androgen hormones in castrated rabbits might contribute to the suppression of the inflammatory response. This notion is in line with the reported difference in response of male and female astrocytes to LPS: the levels of mRNA of pro-inflammatory cytokines were significantly higher in astrocytes derived from males than females (Santos-Galindo *et al.*, 2011).

References

Aydin, M., Dumlu, T., Alemdar, R., Kayapinar, O., Celbek, G., Karabacak, A., Turker, Y., Kaya, H., Ertas, F. and Atilgan, Z., 2012. Correlation between the body fat composition and high sensitive C-reactive protein in Turkish adults. Endocr Regul 46: 147-152.

Costa, L.G., Cole, T.B. and Furlong, C.E., 2005. Paraoxonase (PON1): from toxicology to cardiovascular medicine. Acta Biomed 2: 50-57.

Draganov, D.I. and La Du, B.N., 2004. Pharmacogenetics of paraoxonases: a brief review. Naunyn Schmiedebergs Arch Pharmacol 369: 78-88.

Emin, S., Yordanova, K., Doneva, K., Tsoneva, V., Z., K. and T., V., 2009. Investigation of paraoxonase and arylesterase activity of serum PON1 in helthy individuals. International Scientific Conference, Stara Zagora, June, 4-5, 2009, Proceeding (electron version-CD).

Forouhi, N.G., Sattar, N. and McKeigue, P.M., 2001. Relation of C-reactive protein to body fat distribution and features of the metabolic syndrome in Europeans and South Asians. Int J Obes Relat Metab Disord 25: 1327-1331.

Georgiev, I.P., Georgieva, T.M., Ivanov, V., Dimitrova, S., Kanelov, I., Vlaykova, T., Tanev, S., Zaprianova, D., Dichlianova, E., Penchev, G., Lazarov, L., Vachkova, E. and Roussenov, A., 2011. Effects of castration-induced visceral obesity and antioxidant treatment on lipid profile and insulin sensitivity in New Zealand white rabbits. Res Vet Sci 90: 196-204.

Georgieva, T., Vlaykova, T., Dishlianova, E., Petrov, V. and Penchev, I., 2012. The behaviour of ceruloplasmin as an acute phase protein in obese and infected rabbits and in infected rabbits. In: E.D. Rodrigues P, de Almeida (Ed.), 3rd Manging Committee Meeting and 2nd Meeting of Working groups, 1,2&3 of COST Action FA1002. l, 12-13, April, 2012, Vilamoura, Algarve, Portuga. pp. A67-70.

Lumeng, C.N. and Saltiel, A.R., 2011. Inflammatory links between obesity and metabolic disease. J Clin Invest 121: 2111-2117.

Mackness, M.I., Mackness, B. and Durrington, P.N., 2002. Paraoxonase and coronary heart disease. Atheroscler Suppl 3: 49-55.

Murray, P., Baron, E., Pfaller, M., Tenover, F. and Yolken, R., 1999. Manual of Clinical Microbiology. American Society for Microbiology, Washington, DC.

Nakarai, H., Yamashita, A., Nagayasu, S., Iwashita, M., Kumamoto, S., Ohyama, H., Hata, M., Soga, Y., Kushiyama, A., Asano, T., Abiko, Y. and Nishimura, F., 2012. Adipocyte-macrophage interaction may mediate LPS-induced low-grade inflammation: potential link with metabolic complications. Innate Immun 18: 164-170.

Perry, C.D., Alekel, D.L., Ritland, L.M., Bhupathiraju, S.N., Stewart, J.W., Hanson, L.N., Matvienko, O.A., Kohut, M.L., Reddy, M.B., Van Loan, M.D. and Genschel, U., 2008. Centrally located body fat is related to inflammatory markers in healthy postmenopausal women. Menopause 15: 619-627.

Santos-Galindo, M., Acaz-Fonseca, E., Bellini, M.J. and Garcia-Segura, L.M., 2011. Sex differences in the inflammatory response of primary astrocytes to lipopolysaccharide. Biol Sex Differ 2: 2042-6410.

Tomas, M., Senti, M., Garcia-Faria, F., Vila, J., Torrents, A., Covas, M. and Marrugat, J., 2000. Effect of simvastatin therapy on paraoxonase activity and related lipoproteins in familial hypercholesterolemic patients. Arterioscler Thromb Vasc Biol 20: 2113-2119.

Van Himbergen, T.M., van Tits, L.J., Roest, M. and Stalenhoef, A.F., 2006. The story of PON1: how an organophosphate-hydrolysing enzyme is becoming a player in cardiovascular medicine. Neth J Med 64: 34-38.

Vlaykova, T., Georgieva, T., Dishlyanova, E., Bozakova, N. and Georgiev, I., 2013. Effects of acute Staphylococcus aureus infection on paraoxonase activity, thiol concentrations and ferric reducing ability of plasma in rabbits. Revue de Médecine Vétérinaire (in press).

Part IV
Animal production

Biomarkers of winter disease in gilthead seabream: a proteomics approach

Denise Schrama[1], Nadège Richard[1], Tomé S. Silva[1], Luís E.C. Conceição[1,2], Jorge P. Dias[1,2], David Eckersall[3], Richard Burchmore[3] and Pedro M. Rodrigues[1]

[1]CCMAR, Centro de Ciências do Mar do Algarve, Universidade do Algarve, Campus de Gambelas, 8005-139 Faro, Portugal; pmrodrig@ualg.pt
[2]SPAROS, Centro regional para a inovação do Algarve, Campus de Gambelas, Faro, Portugal
[3]University of Glasgow, Institute of Infection, Immunity and Inflammation, School of Veterinary Medicine, Glasgow, United Kingdom

Introduction

Farmed seafood organisms are susceptible to a wide range of factors that can pose a major threat to a thriving aquaculture industry with considerable economic repercussions. This industry has been going through major challenges in its effort to respond to the growing consumer demand, coupled with clear global market awareness of product quality and animal welfare. A good balance between these challenges may greatly benefit from a better scientific understanding of the biological traits in seafood farming.

In this work we address one of the main problems in farmed species produced in Portugal as well as in the Mediterranean Sea; The Gilthead seabream winter disease syndrome.

Gilthead seabream (*Sparus aurata*), occupies habitats on the east of the Atlantic ocean from the British Isles to Cape Verde, the Canary Islands and in the Mediterranean (Sola, 2005). It is the main fish species being produced in Portugal with worldwide production values reaching 139,000 tons (FAO, 2012) against a total of 8,000 tons in capture in 2010. This species is vulnerable to winter disease, also called winter syndrome. This disease is characterized by a series of metabolic and immune system disorders caused by low temperatures that affect seabream culture (Tort *et al.*, 1998), leading to important industrial losses, either as mortalities or growth decreases during spring.

The mortality rate is around 7-10% but might increase when temperatures are lower (Tort *et al.*, 2004). The first winter period (December-March) influences fish farms in the Northern Mediterranean, where mortalities increase when fish are disturbed. In Portugal the disease appears more frequently at the end of winter time (March-April), also called second winter, during which the pathogen *Pseudomonas anguilliseptica* is commonly isolated (Domenech *et al.*, 1997; Tort *et al.*, 2004). When the syndrome affects the fish, altered swimming patterns are observed and reactivity decreases (Tort *et al.*, 2004). These effects can be minimized through dietary optimization, either improving diets for the end of summer season, or feeding seabream with a fortified diet during autumn/winter. However, the development of a fortified diet for autumn/winter is hindered by the lack of tools to evaluate fish metabolic and physiological condition. Nevertheless, the recent development of new generation molecular tools provides a good perspective to supply molecular indicators of condition to support this work.

A proteomics approach was followed in order to develop in the near future a rapid analytical service that will allow evaluation of the effect of specific ingredients and additives (supplied in a specific diet formulation being developed) on medium/long term immune status and oxidative stress of seabream.

Methodology

This experiment studies the effect of a fortified diet during winter period on gilthead seabream using 2D-DIGE techniques. As a control, a commercial diet was given and 25 fish were kept in duplicate 1000 L tanks in the Ramalhete station of the University of Algarve. For each sampling (one at the end of March and other at the end of June) 10 fish were randomly anesthetized with phenoxyethanol, blood was withdrawn and centrifuged at 5000 g for 20 minutes to collect serum. Proteins were quantified using the Bradford method and 50 µg were labeled with CyDye 3 or 5. CyDye 2 was used as an internal standard. After incubation for half an hour, lysine (10 mM) was used to stop the reaction. Serum proteins were focused on 24 cm Immobiline Drystrips (GE Healthcare, Sweden) with pH 4-7 until a total of 60.000 Vhr. A reduction and alkylation step was done with DTT and iodoacetamide, respectively before separation according to molecular weight on 12.5% polyacrylamide gels. Gels were scanned on a 9400 Typhoon scanner (GE Healthcare, Sweden) and analysed with Decyder 2D version 7.0 (GE Healthcare, Sweden). Spots of interest ($P<0.05$ by ANOVA) were excised manually, trypsinized and sequenced with an LTQ Velos Orbitrap mass spectrometer (Thermo Scientific, USA) followed by a MASCOT search using Actinopterygii as taxonomy.

Results and discussion

A gel obtained from this experiment is represented in Figure 1 with some of the sequenced spots highlighted.

We are currently characterizing all protein spots of interest with several already identified as apolipoprotein A-I – which is the major protein observed in serum of teleosts and contributes in the innate immune system (Concha *et al.*, 2004), and some as transferrin – which is a glycoprotein and binds to iron, being important for the health of teleost (Bury *et al.*, 2003).

Conclusion and future perspectives

Proteins from serum, as well as from liver and head kidney are being further analyzed and eventual biomarkers will be confirmed by western blot using specific antibodies.

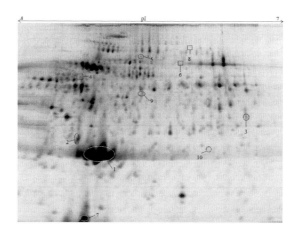

Figure 1. 2D-DIGE gel from a serum sample. Circled spots are up-regulated in control fish and squared spots are up-regulated in fish fed with fortified diet.

Other diets will be developed for a new trial in the winter period, using the commercial diet as control; tissues will be collected at the end of March and June 2013.

An overview of the results obtained from a comparative analysis of the different profiles will provide detailed information regarding the seabream metabolic adjustments to different diets. A first selection of possible biological markers that will allow foreseeing the effect of specific ingredients and additives on medium/long term immune status and oxidative stress of seabream will be available shortly.

Acknowledgements

This work is part of project 21595-INUTR, co-financed by FEDER through PO Algarve 21 in the framework of QREN 2007-2013. D. Schrama acknowledges support of COST Action FA 1002 – Farm Animal Proteomics, through a STSM at the laboratories of Prof. David Eckersall and Richard Burchmore, University of Glasgow.

References

Bury, N.R., Walker, P.A. and Glover, C.N., 2003. Nutritive metal uptake in teleost fish. J Exp Biol 206: 11-23.

Concha, M.I., Smith, V.J., Castro, K., Bastias, A., Romero, A. and Amthauer, R.J., 2004. Apolipoproteins A-I and A-II are potentially important effectors of innate immunity in the teleost fish *Cyprinus carpio*. Eur J Biochem 271: 2984-2990.

Domenech, A., FernandezGarayzabal, J.F., Lawson, P., Garcia, J.A., Cutuli, M.T., Blanco, M., Gibello, A., Moreno, M.A., Collins, M.D. and Dominguez, L., 1997. Winter disease outbreak in sea-bream (*Sparus aurata*) associated with *Pseudomonas anguilliseptica* infection. Aquaculture 156: 317-326.

FAO, 2012. Species fact sheets.

Sola, L., Moretti, A., Crosetti, D., Karaiskou, N., Magoulas, A., Rossi, A.R., Rye, M., Triantafyllidis, A, Tsigenopoulos, C.S., 2005. Gilthead seabream – *Sparus aurata*. Genimpact final scientific report.

Tort, L., Rotllant, J. and Rovira, L., 1998. Immunological suppression in gilthead sea bream *Sparus aurata* of the North-West Mediterranean at low temperatures. Comparative Biochemistry and Physiology a-Molecular and Integrative Physiology 120: 175-179.

Tort, L., Rotllant, J., Liarte, C., Acerete, L., Hernandez, A., Ceulemans, S., Coutteau, P. and Padros, F., 2004. Effects of temperature decrease on feeding rates, immune indicators and histopathological changes of gilthead sea bream *Sparus aurata* fed with an experimental diet. Aquaculture 229: 55-65.

The absorption of colostral proteins in newborn lambs: an iTRAQ proteomics study

Lorenzo E. Hernández-Castellano[1,2], Anastasio Argüello[1], Thomas F. Dyrlund[2], André M. Almeida[3,4], Noemí Castro[1] and Emøke Bendixen[2]
[1]*Department of Animal Science, Universidad de Las Palmas de Gran Canaria, Arucas, Gran Canaria, Spain; lhernandezc@becarios.ulpgc.es*
[2]*Department of Molecular Biology and Genetics, Aarhus University, Aarhus, Denmark*
[3]*Instituto de Tecnología Química e Biologica, Universidade Nova de Lisboa, Oeiras, Portugal*
[4]*Instituto de Investigação Científica Tropical, Lisboa, Portugal*

Introduction

Colostrum is the first source of nutrition in neonatal ruminants, supplying not only nutrients, but having also a fundamental biological function, promoting immunoglobulin transfer from the dam to the newborn. As previously described, newborn ruminants are hypo gammaglobulinemic, as the complexity of the synepitheliochorial ruminant placenta does not allow a sufficient transfer of immunoglobulins (Ig's) from the dam to the foetus (Castro *et al.*, 2011). As a consequence, colostrum intake and colostral protein absorption plays an essential role in Passive Immune Transfer (PIT) and ultimately in the newborn survival rate (Stelwagen *et al.*, 2009).

The aim of this study was to determine which proteins present in the colostrum are absorbed by newborn lambs during the first 14 hours after birth using proteomic methodologies for identification and quantitation. Quantification was based on using the iTRAQ (isobaric tag for relative and absolute quantitation) method. This approach could describe proteinsthat relate to eitherPIT or lamb immune system development.

Material and methods

Sample collection

Eight newborn lambs (Canaria breed) were used for their ability to take up colostral proteins. The experiment took place at the experimental farm of the Veterinary Faculty of the Universidad de Las Palmas de Gran Canaria during the spring 2011. Two groups of animals (of 4 each) were fed colostrum at two different time points after birth. As observed in Figure 1, one group (termed colostrum group; C group) received colostrum, at 2, 14 and 26 hours after birth, while the other group (termed delayed colostrum group; DC group) was fed with colostrum at 14 and 26 hours after birth. At the end of the colostrum feeding period (26 hours after birth), each animal (from both groups) took the same amount of fresh colostrum from a pool with

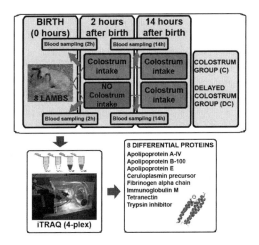

Figure 1. Experimental design and results.

64.74 mg of IgG/ml. Blood samples were collected before feeding at 2 and 14 hours after birth, and the obtained plasma was frozen at -80 °C until further analysis.

Sample preparation

Blood plasma samples were analysed at Aarhus University (Aarhus, Denmark). First, 200 µl of each plasma sample were homogenized with 1 ml of TES buffer (10 mM Tris-HCl (pH 7.6), 1 mM EDTA, 0.25M sucrose) using an ultra-turrax (IKA T110, Staufen, Germany) at 12,000 rpm. After homogenization, samples were centrifuged at 10,000×*g* for 30 min at 4 °C. Protein concentration of the supernatant was determined with the Quick Start™ Bradford Protein Assay (Bio-Rad, Hercules, CA, USA), using BSA as standard reference. After quantitation, 100 µg of protein from each sample were obtained after precipitation with 6 volumes of ice-cold acetone (-20 °C) at 15,000×*g* for 10 minutes at 4 °C.

iTRAQlabeling

Six sets of iTRAQ experiments were design in order to analyse and compare all samples, according to the procedure described by (Danielsen *et al.*, 2010). Consistently, the iTRAQ reagent 114 was used to label a pooled plasma sample consisting of an equal protein amount of each of the 14 samples. The other iTRAQ reagents (115, 116 and 117) were used to label the individual plasma samples from the C and DC groups at 2 and 14 hours after birth. Samples were labelled according to the manufacturer (Applied Biosystems Inc., Foster City, CA). Finally, samples were combined to create 4-plexed samples.

SCX fractionation and LC-MS/MS

Peptide mixtures generated from the digestion of 50 µg of protein were injected into an Agilent 1100 Series capillary HPLC equipped with a Zorbax Bio-SCX Series II, and peptides were eluted with a gradient of increasing NaCl solution. Fractions were collected every minute for 65 minutes and then combined according to their peptide loads into 9-10 pooled samples. The pooled SCX fractions were further separated by a reverse phase liquid chromatography using an Agilent 1100 Series nano-flow HPLC system. LC-MS/MS analyses were performed on a QSTAR Elite mass spectrometer (AB Sciex). The collected MS files were converted to Mascot generic format and used to interrogate an in-house assembled sheep and goat database consisting of sequences from TrEMBL, Swiss-Prot and NCBInr (Updated November 2012, 32,444 sequences) using Mascot 2.3.02 (Matrix Science). The significance threshold (p) was set at 0.05 with an ion score cutoff at 31). Mascot results were parsed using MS Data Miner v.1.1(MDM; Dyrlund *et al.*, 2012) and a final report was generated using MDM, comparing all identified and quantified proteins from the 6 sets of iTRAQ data.

Statistical analysis

Statistical analysis was performed using SAS, Version 9.00 (SAS Institute Inc., Cary, NC). The SAS PROC MIXED procedure for repeated measurements was used to evaluate the effect of colostrum intake (Cgroup vs. DC group) at 2 and 14 hours after birth. A Bonferroni's test was used to evaluate differences between groups.

Results and discussion

Sheep protein databases are still very incomplete, detailed annotation of sheep proteins is hence very limited, even when compared to cattle or goat databases. For this reason, only 99 (6[th]iTRAQ set) to 148 proteins (5[th]iTRAQ set) could be observed and characterized using the shotgun proteome data. From this result, a total of 31 proteins were selected, because they followed a normal distribution in at least 3 of the 4 biological replicates in each of the 4studied groups. A statistical analysis was performed in the selected proteins, as described. A total of 8 proteins were found increased in the Colostrum group at 14 hours after birth. These 8 proteins were identified as shown in Table 1.

It should be highlighted that most of the identified proteins, which were increased in the blood plasma of the colostrum group at 14 hours after birth, play an important role as immunomodulators in different ways during the immune response against infectious disease agents. However, it is important to note that using relative quantification methods like iTRAQ, only proteins present in both groups, and at both studied time points of collection can be included in the comparative analyses. It was therefore not possible to analyse a wide range of proteins that were absent at birth but present in the blood 14 hours after birth due to the

Table 1. Identification and function of colostral proteins overexpressed in the colostrum group (C).

Protein	Function	Reference
Apolipoprotein A-IV	Fat absorption (Intestinal synthesis) Reduces gastric acid secretion Immunomodulatory effect	Vowinkel *et al.*, 2004
Apolipoprotein B-100	Fat absorption Immunomodulatory effect	Peterson *et al.*, 2008
Apolipoprotein E	Fat absorption Immunomodulatory effect	Maezawa *et al.*, 2006
Ceruloplasmin precursor	Iron metabolism Cupper transport	Twomey *et al.*, 2005
Fibrinogen Alpha Chain	Coagulation Increase during acute-phase reaction Promotion of adhesion, migration, chemotaxis and phagocytosis of monocytes and macrophages	Ganheim *et al.*, 2003; Tamzali *et al.*, 2001
Immunoglobulin M	Immune function	Delvin, 2004
Tetranectin	Regulation of plasmin activation Neutrophil migration to the infection	Renckens *et al.*, 2006
Trypsin inhibitor	Reduction of biologically active trypsin Protein structure protection	Rawlings *et al.*, 2004

colostrum intake. For quantifying these proteins, alternative methods must be employed, e.g. Selected Reaction Monitoring (SRM).

Conclusions

In conclusion, early colostrum intake produced a relative increase of eight plasma proteins in the C group. This demonstrates that colostrum is essential not only for its Ig´s content but also for the non-immunoglobulin proteins that play a fundamental role in the activation and attraction of immune cells, the low gastric secretion, among others. The result of this work contributes with information about proteins with immune function that are increased after colostrum intake. These can be used to decrease lamb mortality rates and will also contribute to increase the economic benefit of sheep producers. In the future, further proteomic studies will be necessary, particularly using the SMR approach, in order to increase the general knowledge about the role of colostrum in the PIT.

Acknowledgments

Author Lorenzo Enrique Hernández-Castellano acknowledges financial support from the European Cooperation in Science and Technology (COST) through a STSM from action FA-1002. Author also acknowledges financial support from the Formación del Porfesorado Universitario (FPU) program (Ministry of Education, Madrid, Spain). Author A.M. Almeida acknowledge the support from the program Ciência 2007 from Fundação para a Ciência e a Tecnologia (Lisbon, Portugal).

References

Castro, N., Capote, J., Bruckmaier, R.M. and Arguello, A., 2011. Management effects on colostrogenesis in small ruminants: a review. Journal of Applied Animal Research 39: 85-93.

Danielsen, M., Codrea, M.C., Ingvartsen, K.L., Friggens, N.C., Bendixen, E. and Røntved, C.M., 2010. Quantitative milk proteomics – Host responses to lipopolysaccharide-mediated inflammation of bovine mammary gland. Proteomics 10: 2240-2249.

Delvin, T.M., 2004. Bioquímica. Reverté, Barcelona.

Dyrlund, T.F., Poulsen, E.T., Scavenius, C., Sanggaard, K.W. and Enghild, J.J., 2012. MS Data Miner: A web-based software tool to analyze, compare, and share mass spectrometry protein identifications. Proteomics 12: 2792-2796.

Ganheim, C., Hulten, C., Carlsson, U., Kindahl, H., Niskanen, R. and Waller, K.P., 2003. The acute phase response in calves experimentally infected with bovine viral diarrhoea virus and/or *Mannheimia haemolytica*. J Vet Med B Infect Dis Vet Public Health 50: 183-190.

Maezawa, I., Maeda, N., Montine, T.J. and Montine, K.S., 2006. Apolipoprotein E-specific innate immune response in astrocytes from targeted replacement mice. Journal of Neuroinflammation 3.

Peterson, M.M., Mack, J.L., Hall, P.R., Alsup, A.A., Alexander, S.M., Sully, E.K., Sawires, Y.S., Cheung, A.L., Otto, M. and Gresham, H.D., 2008. Apolipoprotein B Is an Innate Barrier against Invasive *Staphylococcus aureus* Infection. Cell Host & Microbe 4: 555-566.

Rawlings, N.D., Tolle, D.P. and Barrett, A.J., 2004. Evolutionary families of peptidase inhibitors. Biochemical Journal 378: 705-716.

Renckens, R., Roelofs, J.J.T.H., Florquin, S. and van der Poll, T., 2006. Urokinase-type plasminogen activator receptor plays a role in neutrophil migration during lipopolysaccharide-induced peritoneal inflammation but not during *Escherichia coli*-induced peritonitis. Journal of Infectious Diseases 193: 522-530.

Stelwagen, K., Carpenter, E., Haigh, B., Hodgkinson, A. and Wheeler, T.T., 2009. Immune components of bovine colostrum and milk. J Anim Sci 87: 3-9.

Tamzali, Y., Guelfi, J.F. and Braun, J.P., 2001. Plasma fibrinogen measurement in the horse: comparison of Millar's technique with a chronometric technique and the QBC-Vet Autoreader. Res Vet Sci 71: 213-217.

Twomey, P.J., Vijoen, A., House, I.M., Reynolds, T.M. and Wierzbicki, A.S., 2005. Relationship between serum copper, ceruloplasmin, and non-ceruloplasmin-bound copper in routine clinical practice. Clinical Chemistry 51: 1558-1559.

Vowinkel, T., Mori, M., Krieglstein, C.F., Russell, J., Saijo, F., Bharwani, S., Turnage, R.H., Davidson, W.S., Tso, P., Granger, D.N. and Kalogeris, T.J., 2004. Apolipoprotein A-IV inhibits experimental colitis. J Clin Invest 114: 260-269.

Plasma proteome profiles predict diet-induced metabolic syndrome and the early onset of metabolic syndrome in a pig model

Marinus F.W. te Pas[1], Sietse-Jan Koopmans[2], Leo Kruijt[1] and Mari A. Smits[1]
[1]Animal Breeding and Genomics Centre (ABGC), Wageningen UR Livestock Research, Lelystad, the Netherlands; marinus.tepas@wur.nl
[2]Department of Animal Sciences, Adaptation Physiology Group of Wageningen University, P.O. Box 338, 6700 AH Wageningen, the Netherlands

Objectives

Obesity and related diabetes are important health threatening multifactorial metabolic diseases and it has been suggested that 25% of all diabetic patients are unaware of their patho-physiological condition. Feeding behavior is often associated with the onset of the metabolic syndrome. We have developed a pig model for the study of metabolic syndrome and diabetes in Yorkshire × Landrace pigs. We used pigs because of human and pigs share many physiological similarities. The aim is to develop biomarkers for the early detection of the first signs of metabolic changes related to metabolic syndromes like diabetes. The objective of this investigation was to develop proteomics profiles differentiating healthy pigs and pigs with signs of early onset of metabolic syndromes, and with related insulin resistance metabolic syndromes. The results of this study suggest that metabolomics profiling of blood may be a powerful tool for measuring/monitoring physiologic conditions and/or phenotypic traits of pigs.

Material and methods

Animal model and feeding conditions

The performed research is in compliance with the ARRIVE guidelines on animal research. Experimental protocols describing the management, surgical procedures, and animal care were reviewed and approved by the ASG-Lelystad Animal Care and Use Committee (Lelystad, The Netherlands).

Twenty-four 11 weeks old male Yorkshire × Landrace pigs were divided in a control (n=11) and an experimental (n=13) group. The experimental group received Streptozotocin that induces diabetes by destroying the B-cells in the pancreas (Koopmans *et al.*, 2006; Van den Heuvel *et al.*, 2012). Half of the animals in both groups received a 'cafeteria diet' consisting of high saturated fat, half of the animals received a 'Mediterranean diet' consisting of high unsaturated fat for ten weeks. A number of physiological parameters related to body weight gain, body composition, fatness metabolism, and diabetes were measured in order to determine the physiological effects of the different treatments and feeds given to the animals.

Proteomics

To fractionate the plasma proteome isolates we used protein array types H50 (a hydrophobic protein array type for lipoprotein binding with 10% acetonitrile, 0.1% trifluoroacetic acid binding buffer), IMAC (a copper containing protein array type for phosphoprotein binding with 0.1 M sodium phosphate, 0.5 M NaCl pH 7 binding buffer), and CM10 (a weak cation exchanger with low stringent (100 mM sodium-acetate pH 4) and high stringent (100 mM sodium-acetate pH 7) binding buffers for binding of electrically charged proteins). Mass analysis was performed using the SELDI-TOF (Surface-enhanced laser desorption/ionization-time of flight mass spectrometry) (BioRad, Veenendaal, The Netherlands) technology with the Protein Chip System Series 4000 (BioRad) equipment in positive operating mode and laser energy of 3000nJ (Mach *et al.*, 2010).

Statistical evaluation

The proteomics data were analyzed using Lucid Proteomics Software v2.0 (Bio-Rad). The results were evaluated for differential expression between all four groups of pigs. The software provided for the statistical evaluation tools. Significance levels were set to $P<0.005$. Correlations between peak expression levels and metabolic parameters were determined with Spearman correlation.

Results and discussion

Proteomics profiles differentiating control and diabetic pigs fed the Mediterranean or cafeteria diet

We observed a total of 984 protein peaks in the three different protein arrays. The number of peaks was approximately similar for all protein arrays and testing conditions. The number of peaks per testing conditions varied between 6 and 29.

Individual protein peaks differed for their expression levels between the experimental groups. The results showed that in both the control and diabetic animals given the Mediterranean diet the expression level of the protein was similar with little variation between the animals. Cafeteria feed changed the expression level of a number of peaks in both the control and diabetic pigs, but the diabetic pigs showed larger physiological reaction effects than the control pigs. Ten physiological parameters showed correlations with peak expression levels >0.85 (Table 1). These include the parameter measurements of NEFA, triglycerides, glucose levels, and cholesterol, LDL and VLDL, the latter three showed correlations >0.9. These proteome peak profiles can be regarded as potential new biomarkers for the clinical diabetes disease.

Table 1. Number of peaks showing correlation>0.85 between protein peak expression levels and physiological parameters in control and diabetic pigs fed with the Mediterranean or cafeteria diets.

Parameter[1]	N peak correlation control and diabetic pigs	N peak correlation control pigs
Cholesterol	25	1
VLDL	27	2
LDL	26	3
Triglycerids (TG)	1	
TG postprandial	1	
TG in Kidney	1	
NEFA	1	
NEFA postprandial	2	2
Glucose	1	1
Glucose postprandial	2	3
Insulin postprandial		4
Abdominal fat depot (i.e. Omental fat) (g)		2
Omental fat depot per kg body fat		2
Retroperitoneal fat (g)		1
Retroperitoneal fat depot per kg body weight		1
Body weight at slaughter		4

[1] VLDL = Very Low Density Lipoproteins; LDL = Low Density Lipoproteins; NEFA = Non Esterified Fatty Acids.

The protein peaks showed differential expression in at least one of the four groups of pigs. Therefore, these profiles may be also potential biomarkers to detect early onset of the metabolic syndrome. Two main differences were found: (1) peaks showing differential expression between control and diabetic induced pigs fed with the Mediterranean diet; and (2) peaks showing differential expression between control and diabetic induced pigs fed with the cafeteria feed, with the latter group showing larger differences in peak expression levels. This may indicate that diabetic pigs are extremely sensitive to differences in feed composition. These pigs may be metabolically less efficient for fat metabolism.

Dietary effects in control pigs

Analyzing the effects of the diet in the control pigs only may give insight in the early stages of metabolic disease. Peaks with high correlations were found for fat metabolism, glucose / insulin related parameters and for body weight (Table 1). Two-group comparison comparing the Mediterranean and cafeteria dietary effects in the control pigs showed that the saturated fat feed affected the expression of protein peaks related to the cholesterol, LDL, and VLDL groups, although to lower levels compared to the four – group analysis.

These correlations suggest that the diet indeed induced early stages of metabolic syndrome. The blood glucose levels showed that these pigs were not diabetic (yet), so this suggest that we have here the early signs of developing metabolic syndrome, e.g. diabetes. The correlated protein peaks may be valuable indicators of early onset of diabetes and could be useful for the development of biomarkers to detect early onset of diabetes. The results of this study suggest that metabolomics profiling of blood may be a powerful tool for measuring/monitoring physiologic conditions and/or phenotypic traits of pigs.

Acknowledgements

The proteomics work was funded by the Dutch ministry EL&I, grant KB 15-011-003-ASG-V-1.

References

Koopmans S-J, Mroz Z, Dekker R, Corbijn H, Ackermans M, and Sauerwein H, 2006. Association of insulin resistance with hyperglycemia in streptozotocindiabetic pigs. Effects of metformin at soenergetic feeding in a type 2-like diabetic pig model. Metabolism 55: 960-971.

Mach N, Keuning E, Kruijt L, Hortós M, Arnau J, and te Pas MFW, 2010. Comparative proteomic profiling of two muscles from five divergent pig breeds using SELDI-TOF proteomics technology. J Anim Sci 88: 1522-1534.

Van den Heuvel M, Sorop O, Koopmans S-J, Dekker R, de Vries R, van Beusekom HMM, Eringa EC, Duncker DJ, Danser AHJ, and van der Giessen WJ, 2012. Coronary microvascular dysfunction in a porcine model of early atherosclerosis and diabetes. Am J Physiol Heart Circ Physiol 302: H85-H94.

Changes in protein abundance in the vertebral columnof Atlantic salmon (*Salmosalar*) fed variable dietary P levels

Eva Veiseth-Kent[1], Mona E. Pedersen[1], Kristin Hollung[1], Elisabeth Ytteborg[1], Grete Bæverfjord[1], Harald Takle[1], Torbjørn E. Åsgård[1], Robin Ørnsrud[2], Erik-Jan Lock[2] and Sissel Albrektsen[1]

[1]*Nofima – Norwegian Institute of Food, Fisheries and Aquaculture Research, Norway; eva.veiseth-kent@nofima.no*
[2]*NIFES – National Institute of Nutrition and Seafood Research, Norway*

Objectives

Previously we have applied a comparative 2DE-approach to investigate the proteome of Atlantic salmon (*Salmo salar*) with vertebral deformities (Pedersen *et al.*, 2011). This work lead to the identification of a new interesting candidate for the process that leads to fused vertebrae; matrilin-1. This finding was further confirmed by real-time PCR analysis and immunohistochemistry, and illustrates the strength of applying a non-hypothetical approach towards identifying new markers for bone development.

In this study, we therefore chose the same non-hypothetical approach to assess if the protein abundance in vertebral columns of seawater transferred under-yearling Atlantic salmon changes as a response to receiving diets with sub-optimal or adequate dietary phosphorus hydrolysed fromblue whiting (*Micromesistius poutassou*) fish bone meal.

Material and methods

Underyearling Atlantic salmon smolt (Salmo breed) obtained from a commercial hatchery was transported to Nofima's aquaculture research station in Austevoll, Bergen. Following a 6-week acclimation period, the salmon (~210 g) were randomly distributed to six 2×2 m glass fibre tanks, each holding 60 fish, and fed either a control or an experimental dietfor a period of 13 weeks. The main difference between these two diets was the presence of fish bone meal hydrolysate providing additional dietary P to the experimental diet. The marine phosphorus source was added to a concentration of 45 g/kg feed in the experimental diet. The following water parameters were kept during the experiment: temperature of 9.1 °C, salinity of 3.1-3.2%, water flow of 50-55 l/min, and the oxygen of the outlet water not lower than 7.5 mg/l (85% saturation). The fish was exposed to 24 hours light during the experimental period of 91 feeding days. Weight and length were recorded on individual fish from each tank at the start and end of the experiment. All sampled fish were anaesthetised with benzocaine (25 µg/kg), and killed by a blow to the head. The vertebral region below the dorsal fin wascarefully dissected from three fish from each tank (i.e. 9 fish per diet), and snap frozen in liquid nitrogen and stored at -80 °C until further analysis.

The vertebral samples were pulverised in liquid nitrogen before total RNA was isolated using Trizol® (Invitrogen, USA) and reagents from the RNAeasy Micro to Midi Kit® (Quiagen, Germany). Proteins were isolated from the residual Trizol protein fraction following RNA extraction according to the manufacturer's protocol. The protein pellets were solubilised in sample buffer (7 M urea, 2 M thiourea, 2% CHAPS, 1% DTT, 0.5% IPG buffer pH 3-10), and protein concentrations were measured with a commercial kit at 750 nm (RC DC Protein Assay, Bio-Rad, USA) in a spectrophotometer (Ultrospec 3000, Pharmacia Biotech, Sweden) using BSAas a standard. Fifty micrograms of protein extracted from each fish were labelled with CyDye DIGE Fluor minimal dyes (GE Healthcare, USA) according to the manufacturer's protocol, and each sample was labelled with both Cy3 and Cy5 as technical replicates and run on separate gels. In addition, a pooled sample, which served as an internal control, was labelled with Cy2 and included on each gel. The differentially labelled samples were then mixed and loaded onto IPG strips(24 cm, pH 5-8) by in-gel rehydration overnight. Following IEF and SDS-PAGE,the gels were scanned directly using the Ettan DIGE Imager (GE Healthcare) and imported to ProgenesisSameSpots v 4.1 for spot detection and alignment. Artifacts were removed manually before spot detection and statistical analysis.

Significantly altered protein spots were excised from preparative 2-DE gels for trypsin treatment and peptide extraction, and the resulting peptide mixtures were desalted and concentrated using small disks of C18 Empore Disks (3M, USA). Peptides were eluted with 0.8 μl matrix solution (α-cyano-4-hydrozy-cinnamic acid (BrukerDaltonics, Germany) saturated in a 1:1 solution of ACN and 0.1% TFA) and spotted directly onto the MALDI-TOF target plate. An Ultraflex MALDI-TOF/TOF mass spectrometer with a LIFT module (BrukerDaltonics) was used for mass analyses of the peptide mixtures. FlexAnalysis(Version 3.3, BrukerDaltonics) was used to create the peak lists, and BioTools (Version 3.2, BrukerDaltonics) was used for interpretation of MS and MS/MS spectra. Proteins were identified by peptide mass fingerprinting using the database search program MASCOT (http://www.matrixscience.com).

Results and discussion

The statistical analysis of the 2DE-DIGE data showed that 25 spots were significantly altered in the vertebral column of fish fed the experimental diet compared to the control (Figure 1). All but two of these showed a reduced abundance in the vertebral column of fish fed the experimental diet. Eleven of these affected proteins were successfully identified (Table 1). The annexin proteins have been found to play important roles in osteoblast differentiation and mineralisation (Damazo et al., 2007; Gillette and Nielsen-Preiss, 2004), and similar functions have also been observed formimecan (Bentz et al., 1989; Rehn et al., 2008). Moreover, both assembly and disassembly of actin filaments have been reported to be enhanced during the early stages of osteoblast differentiation (Hong et al., 2010), and we observed changes in both actin and a capping protein that regulates actin filament dynamics. Finally, serotransferrin is involved in transport of iron, and it is well known that iron is required for normal osteoblast development (Messer et al., 2010).

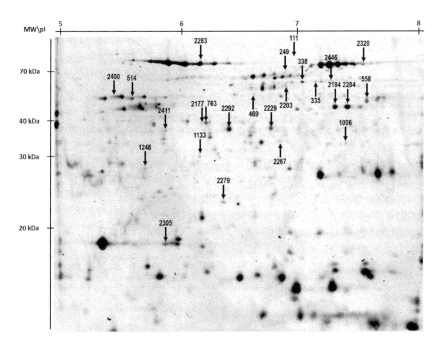

Figure 1. 2DE gel of proteins extracted from the vertebral column of Atlantic salmon. A representative 2-DE gel of proteins extracted from the vertebral column of Atlantic salmon. Protein spots that were found to change in abundance between the two feeding groups are numbered.

Table 1. Proteins from the vertebral column of Atlantic salmon changed in abundance between the groups fed control and experimental diet.

Spot no.	Protein name (source)	NCBI ac. no.	Match pep./ % seq. cov.	Ratio exp: con
514	Enolase 3-1 (*Salmo salar*)	gi|213511756	12 / 36	0.62
763	Annexin A2-A (*Salmo salar*)	gi|213511977	15 / 39	0.58
1006	Mimecan precursor (*Salmo salar*)	gi|221219504	10 / 34	0.42
1133	Mimecan precursor (*Salmo salar*)	gi|213513034	8 / 27	0.73
2184	Beta actin (*Hippoglossus hippoglossus*)	gi|45237481	11 / 32	0.54
2267	Capping protein (actin filament) muscle Z-line beta (*Salmo salar*)	gi|197631853	11 / 35	0.73
2283	Serotransferrin-1 precursor (*Salmo salar*)	gi|185132395	16 / 22	0.64
2284	Fast skeletalmusclealpha-actin (*Gadus morhua*)	gi|22654302	11 / 35	0.57
2292	Annexin A1 (*Salmo salar*)	gi|213510942	16 / 48	0.62
2305	Non-metastatic cells 1 protein (*Salmo salar*)	gi|213510934	10 /51	0.65
2411	Mimecan precursor (*Salmo salar*)	gi|213513034	9 / 30	0.68

In conclusion, the observed reduction in abundance of these proteins in fish fed the experimental diet may indicate changes in mineralisation and development of osteoblasts in their vertebrae. However, interpretation of the present results is complex, and it is necessary to perform further analyses including histological studies in order to evaluate possible impacts on vertebral development.

Acknowledgements

Financial support for this study was provided byThe Norwegian Seafood Research Fund (FHF).

References

Bentz, H., Nathan, R.M., Rosen, D.M., Armstrong, R.M., Thompson, A.Y., Segarini, P.R., Mathews, M.C., Dasch, J.R., Piez, K.A. and Seyedin, S.M., 1989. Purification and characterization of a unique osteoinductive factor from bovine bone. J Biol Chem 264: 20805-20810.

Damazo, A.S., Moradi-Bidhendi, N., Oliani, S.M. and Flower, R.J., 2007. Role of annexin 1 gene expression in mouse craniofacial bone development. Birth Defects Res A Clin Mol Teratol 79: 524-532.

Gillette, J.M. and Nielsen-Preiss, S.M., 2004. The role of annexin 2 in osteoblastic mineralization. J Cell Sci 117: 441-449.

Hong, D., Chen, H.X., Yu, H.Q., Liang, Y., Wang, C., Lian, Q.Q., Deng, H.T. and Ge, R.S., 2010. Morphological and proteomic analysis of early stage of osteoblast differentiation in osteoblastic progenitor cells. Experimental Cell Research 316: 2291-2300.

Messer, J.G., Cooney, P.T. and Kipp, D.E., 2010. Iron chelator deferoxamine alters iron-regulatory genes and proteins and suppresses osteoblast phenotype in fetal rat calvaria cells. Bone 46: 1408-1415.

Pedersen, M.E., Takle, H., Ytteborg, E., Veiseth-Kent, E., Enersen, G., Faergestad, E., Baeverfjord, G. and Hannesson, K.O., 2011. Matrilin-1 expression is increased in the vertebral column of Atlantic salmon (*Salmo salar* L.) individuals displaying spinal fusions. Fish Physiology and Biochemistry 37: 821-831.

Rehn, A.P., Cerny, R., Sugars, R.V., Kaukua, N. and Wendel, M., 2008. Osteoadherin is upregulated by mature osteoblasts and enhances their *in vitro* differentiation and mineralization. Calcified Tissue International 82: 454-464.

Neurotransmitter levels and proteomic approach in pig brain: pre-slaughter handling stress and cognitive biases

Laura Arroyo[1], Anna Marco-Ramell[1], Miriam Soler[1], Raquel Peña[1], Antonio Velarde[3], Josefina Sabrià[2], Mercedes Unzeta[2] and Anna Bassols[1]
[1]*Dept. Bioquímica, Universitat Autònoma de Barcelona, Spain; arroyo.sanchez.l@gmail.com*
[2]*Institute of Neurosciences, Universitat Autònoma de Barcelona, Spain*
[3]*IRTA, Monells, Girona, Spain*

Introduction

Stress can be defined as a brain-body reaction towards stimuli arising from the environment or from internal cues that are interpreted as a disruption of homeostasis. The organization of the response to a stressful situation involves the activity of several areas of the limbic system through neurotransmitters synthesis (Mora *et al.*, 2012). Changes in a neurotransmitter's concentrations are also related to the activation and modulation of behavioral processes and autonomic response (Herman *et al.* 2005).

Moreover, the development of methods for assessing the affective (or emotional) states is a crucial step in improving animal welfare. The 'cognitive bias', defined as a pattern of deviation in judgment in particular situations, is used as a marker (optimistic or pessimistic) for the effects of the affective state on cognitive processes (Douglas *et al.*, 2012).

Objectives

The aim of this study was to determine the concentration of indoleamines and catecholamines in the hypothalamus, amygdala, hippocampus and prefrontal cortex of a group of slaughtered pigs exposed to handling stress, classified according their emotional state.

In addition, we designed a proteomic approach to study brain tissue to identify proteins expressed in these areas.

Material and methods

The study was carried out on 36 hybrid Large White × Landrace male pigs housed at Institut de Recerca i Tecnologia Agroalimentàries (IRTA)-Monells facilities. Animals were trained to learn to discriminate positive and negative spatial cues. They were classified according to their behavior (optimistic or pessimistic emotional state) during the cognitive bias test (Douglas *et al.*, 2012), when an ambiguous cue was presented before they were subjected to stress handling at the slaughterhouse.

Animals were subjected to stress handling in the slaughterhouse. The brain was quickly removed and the selected structures -hypothalamus, hippocampus, amygdale and prefrontal cortex- were excised, dissected, frozen in liquid nitrogen and stored at -80 °C.

Before analysis the samples were weighted and homogenized (1:10 w/v) in ice-cold homogenization buffer (0.250 mM $HClO_4$, 0.100 mM $Na_2S_2O_5$, 0.250 mM EDTA). After centrifugation, the supernatant was used to determine the concentration of noradrenaline (NA), dopamine (DA), 3,4.dihydroxyphenylacetic acid (DOPAC), homavanillic acid (HVA), 5-hydroxyindole-3-acetic acid (5-HIAA) and serotonin (5-HT) using a high-performance liquid chromatography (HPLC) with electrochemical detection.

In addition, 70 mg prefrontal cortex tissue was mixed with 0.6 ml lysis buffer (30 mM Tris-HCl, 7 M urea, 2 M thiourea, 4% CHAPS, Protease Inhibitor Cocktail (Sigma), pH 8) and dessalted. The isoelectrofocusing was performed with 100 µg prefrontal cortex protein on 7 cm pH 3-10 immobilized IPG strips (GE Healthcare). Then, proteins were separated by molecular weight in a 12% SDS-PAGE gel and stained with silver.

Results and discussion

Catecholamines and indoleamines determination

Regional distribution of brain monoamines showed similar patterns to those described in the literature (Piekarzewska *et al.*, 2000). However, animals defined in a positive emotional state showed higher concentrations of indoleamines and dopamine in the prefrontal cortex ($P<0.05$) and lower concentrations of NA and 5-HT in the hippocampus ($P<0.1$), suggesting a relation with motivation and cognition (Table 1).

On the other hand, handling stress produced changes in the concentrations of catecholamines and indoleamines in the amygdala and the hippocampus. As compared to unstressed controls, catecholamines and serotonin levels in the amygdale were lower in pigs exposed to handling stress ($P<0.05$) whereas hippocampal levels of indoleamines were increased ($P<0.05$) (Table 2). This suggests that the amygdala, involved in emotion modulation and cognition, plays a role in the stress response.

2-DE approach

We optimized the conditions for sample preparation and isoelectrofocusing when working with brain tissue, as seen in Figure 1.

Table 1. Neurotransmitter mean concentrations (ng/g tissue) in the amygdala, prefrontal cortex, hippocampus and hypothalamus in pessimistic (Pessim.) and optimistic (Optim.) pigs.

	Amygdala		Prefrontal cortex		Hippocampus		Hypothalamus	
	Pessim.	Optim.	Pessim.	Optim.	Pessim.	Optim.	Pessim.	Optim.
NA	150.56	164.36	99.75	82.25	151.85[b]	123.65[b]	2,130.24	1,961.30
DOPAC	41.54	43.13	0.00	1.61	0.00	0.00	54.55	66.52
DA	417.64	358.11	10.26[a]	38.02[a]	17.79	26.20	277.52	354.79
HVA	347.92	278.80	41.55	30.25	8.87	9.87	379.41	340.30
5-HIAA	251.38	275.48	74.57	73.36	137.25	125.34	408.14	371.63
5-HT	880.95	938.08	223.83[a]	258.72[a]	320.90[a]	262.58[a]	1068.37	997.96
CATECHOLAMINES	807.09	680.03	51.81	69.88	26.65	36.15	1,006.67	1,037.08
INDOLEAMINES	1,132.33	1,213.56	298.40[a]	332.07[a]	458.16[b]	387.92[b]	1,476.51	1,369.59

[a] Significant differences between groups ($P<0.05$).
[b] Significant differences between groups ($P<0.1$).

Table 2. Neurotransmitter mean concentrations (ng/g tissue) in the amygdala, prefrontal cortex, hippocampus and hypothalamus in control and stress-exposed pigs.

	Amygdala		Prefrontal cortex		Hippocampus		Hypothalamus	
	Control	Stress	Control	Stress	Control	Stress	Control	Stress
NA	146.79	158.55	80.39	90.19	134.21	126.75	2,116.99	1,829.08
DOPAC	41.54	43.13	0.00	1.61	0.00	0.00	54.55	66.52
DA	403.52[b]	313.27[b]	30.51	30.14	30.23	26.79	293.10	353.59
HVA	318.94[a]	239.32[a]	36.69	41.84	7.26	16.20	269.01	392.65
5-HIAA	292.67[b]	248.00[b]	74.10	73.67	118.24[a]	141.23[a]	352.19	390.07
5-HT	1,009.16[a]	820.20[a]	244.27	261.49	257.78[b]	295.35[b]	967.30	1,018.00
CATECHOLAMINES	769.84[a]	587.15[a]	68.19	72.90	37.70	42.99	908.33	1,045.22
INDOLEAMINES	1,301.83[a]	1,068.20[a]	318.37	335.16	376.02[a]	436.58[a]	1,319.50	1,408.07

[a] Significant differences between groups ($P<0.05$).
[b] Significant differences between groups ($P<0.1$).

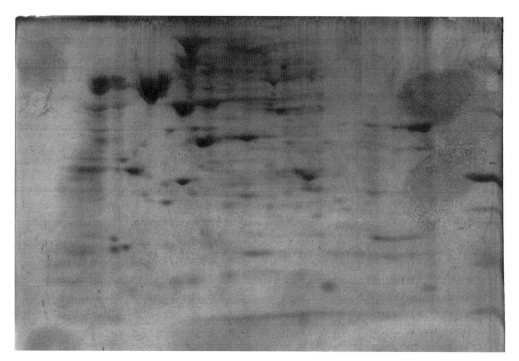

Figure 1. Silver-stained 2-DE gel from prefrontal cortex tissue.

Conclusions

We conclude that stress produces changes in the catecholamine and indoleamine synthesis in the limbic system, especially in the amygdale and hippocampus. Furthermore, we also found differences in neurotransmitter's concentration in the hippocampus and prefrontal cortex between animals defined as optimistic or pessimistic.

Furthermore, we confirm proteomics as a tool for protein screening in brain tissues.

References

Douglas, C., Bateson, M., Walsh, C., Bédué, A. and Edwards, S.A., 2012. Environmental enrichment induces optimistic cognitive biases in pigs. Applied Animal Behaviour Science 139: 65-73.

Herman, J.P., Ostrander, M.M., Mueller, N.K. and Figueiredo, H., 2005. Limbic system mechanisms of stress regulation: Hypothalamo-pituitary-adrenocortical axis. Progress in Neuro-Psychopharmacology and Biological Psychiatry 29: 1201-1213.

Mora, F., Segovia, G., del Arco, A., de Blas, M. and Garrido, P., 2012. Stress, neurotransmitters, corticosterone and body-brain integration. Brain Research 1476: 71-85.

Piekarzewska, A.B., Rosochacki, S.J. and Sender, G., 2000. The Effect of Acute Restraint Stress on Regional Brain Neurotransmitter Levels in Stress-Susceptible Pietrain Pigs. Journal of Veterinary Medicine Series A 47: 257-269.

The sheep (*Ovis aries*) mammary gland mitochondrial complexes: establishment of a Blue-Native PAGE separation method as a model for other ruminant species

Carlos Filipe[1], Vitor Vasconcelos[1,2], André M. Almeida[2,3] and Alexandre Campos[1]

[1]*CIMAR/CIIMAR, Centro Interdisciplinar de Investigação Marinha e Ambiental, Universidade do Porto, Porto, Portugal; acampos@ciimar.up.pt*

[2]*Departamento de Biologia, Faculdade de Ciências,Universidade do Porto, Porto, Portugal*

[3]*Instituto de Investigação Científica Tropical, Lisboa, Portugal and CIISA-Centro, Interdisciplinar de Investigação em Sanidade Animal, 1300, Lisboa, Portugal*

Objectives

Seasonal weight loss (SWL) is the most important limitation to animal production in tropical and Mediterranean regions, conditioning commercial producer's incomes and the nutritional status of rural communities (Almeida *et al.*, 2006; Almeida and Cardoso, 2008). It is of outmost importance to produce strategies to oppose adverse effects of SWL. Breeds that have evolved in harsh climates have acquired a tolerance to SWL through selection. Most of the factors determining such ability are related to biochemical metabolic pathways and are likely important biomarkers to SWL. The objective of the present work is to establish a Blue-Native protocol to separate and study mitochondrial protein complexes isolated from mammary glands, as a tool to investigate molecular markers of tolerance to SWL in sheep (*Ovis aries*) as a model for other dairy ruminants, particularly goats. After mitochondrial isolation, protein complexes were solubilized with different detergent concentrations. The protein profiles were thereafter assessed by BN-PAGE.

Material and methods

Sheep mammary-gland samples were obtained from Portuguese white merino ewes at the Veterinary Medicine Faculty (Lisbon, Portugal). Animals were euthanized following standard procedures and under competent veterinary supervision. Samples (secretory tissue) were dissected with scalpel and immediately stored at -80 °C until further analysis. The tissues were disrupted in liquid nitrogen and homogenized in sucrose (0.35 M), EDTA (1.5 M) Tris (1 mM) and BSA (1%, w/v), pH 7.4 (Mehard *et al.*, 1971). Additional homogenization was accomplished by sonication, 3 cycles at 60 Hz during 5s (VibraCell 50-sonics & Material Inc. Danbury, CT, USA). Mitochondrial fractions were subsequently obtained by differential centrifugation. The homogenate was first centrifuged for 10 min, at 700 g and 4 °C to discard cell debris. A second centrifugation at 8,800×*g*, 10 min was performed and the mitochondria pellet retained. The organelle was subsequently washed with aminocaproic acid (ACA) (750 mM), BisTris (50 mM), Na-EDTA (0.5 mM), pH 7.0 (BN sample buffer), centrifuged and finally suspended to a concentration of 0.26 mg/ml FW mitochondria (Reisinger and Eichacker, 2008). Total protein was quantified with Bradford method and protein complexes solubilized with the detergents triton x-100 and n-Dodecyl-beta-D-maltoside (DDM). The following protein:detergent (w/w) ratios were attempted to test protein solubilisation: 1:2.5, 1.5, 1:3.5, 1:5, 1:7. After detergent treatment, coomassie blue G 250 (5%, w/v) and ACA (750 mM) solution was added to the samples (1/10 of sample volume). Coomassie blue will bind to proteins enabling a negative charge for the separation by electrophoresis (Reisinger and Eichacker, 2008). Mitochondrial protein complexes were thereafter separated by Blue-Native using homogeneous polyacrylamide gels (6.5% and 7.5% acrylamide) (Reisinger and Eichacker, 2008). Protein bands were visualized with silver nitrate (Blum *et al.*, 1987).

Results and discussion

Membrane protein solubilisation is regarded as one of the critical steps in BN-PAGE analysis (Reisinger and Eichacker, 2008). It is therefore suggested, as a starting point of a proteomics investigation, to undertake an optimization of this step, exploring for instance, the properties and concentrations of detergents. In this study mitochondrial protein complexes from ewe mammary gland tissues were solubilized with the non-denaturing detergents triton-x and DDM and using the following protein:detergent ratios (w/w): 1:2.5, 1.5, 1:3.5, 1:5, 1:7. The protocol used to extract mitochondria from mammary gland tissues enabled us to obtain samples with relatively low total protein (0.5 mg/ml prot), nevertheless upon BN-PAGE a total of 7 and 8 protein complexes were detected in triton-x and DDM treated mitochondrial samples (Figure 1). Protein complexes with molecular mass between 242 and 720 kDa or higher were separated in BN-PAGE homogeneous gels (Figure 1). The higher number of bands and enhanced band intensity retrieved by DDM indicates that this detergent is more efficient to solubilize membrane proteins. Furthermore increasing protein:detergent ratio from 1:2.5 to 1:7.0 did not improve protein solubilisation, i.e. in what regards protein band number and intensity. The results are consistent, for example, with the separation of mitochondrial membrane complexes I-V accomplished in different rat organs (Reifschneider *et al.*, 2008).

This study is a first attempt to perform BN-PAGE in sheep mammary gland tissues. Further BN-PAGE studies are foreseen to establish this technique including testing the detergent digitonin, performing a second-dimension in denaturing conditions and assessment of the function of the distinct protein complexes by mass spectrometry. The protocol herein described is of use in other dairy species such as cattle and goat.

Acknowledgements

This research was supported by the research project PTDC/CVT/116499/2010 – Lactation and milk production in goat (*Capra hircus*): identifying molecular markers underlying adaptation to seasonal weight loss from Fundação para a Ciência e Tecnologia (FCT, Lisbon, Portugal) and by the European Regional Development Fund (ERDF) through the COMPETE – Operational

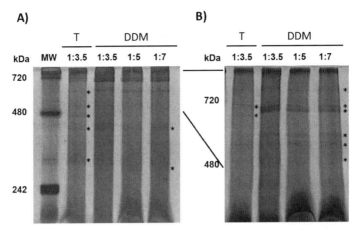

Figure 1. BN-PAGE of mitochondrial protein complexes from ewe mammary gland tissues. Protein complexes separated in 7.5% (A) and 6.5% (B) acrylamide gels. Mitochondrial samples treated with triton X-100 (T) and n-Dodecyl-beta-D-maltoside (DDM). On the top of each lane are the corresponding protein:detergent (w:w) ratios used to solubilize protein complexes. Protein bands detected with silver stain (*).

Competitiveness Programme and national funds through FCT – Foundation for Science and Technology, under the project 'PEst-C/MAR/LA0015/2011'. A Campos and AM Almeida contract work is supported by the Ciência 2007 program by FCT.

References

Almeida, A.M. and Cardoso, L.A., 2008. Animal Production and Genetic Resources in Guinea Bissau: I – Northern Cacheu Province. Tropical Animal Health and Production 40: 529-536.

Almeida, A.M., Schwalbach, L.M., de Waal, H.O., Greyling, J.P. and Cardoso, L.A., 2006. The effect of supplementation on productive performance of Boer goat bucks fed winter veld hay. Tropical Animal Health and Production 38: 443-449.

Blum, H., Beier, H. and Gross, H.J., 1987. Improved silver staining of plant proteins, RNA and DNA in polyacrylamide gels. Electrophoresis, 8:93-99.

Mehard, C.W., Packer, L. and Abraham, S., 1971. Activity and ultrastructure of mitochondria from mouse mammary gland and mammary adenocarcinoma. Cancer Res 31:2148-2160.

Reifschneider, N.H., Goto, S., Nakamoto, H., Takahashi, R., Sugawa, M., Dencher, N.A., Krause, F., 2006. Defining the mitochondrial proteomes from five rat organs in a physiologically significant context using 2D blue-native/SDS-PAGE. J Proteome Res 5:1117-1132.

Reisinger, V. and Eichacker, L.A., 2008. Solubilization of membrane protein complexes for blue native PAGE. Journal of Proteomics,71:277-283.

Bitter taste in water-buffalo (*Bubalus bubalis*): from T2R gene identification to expression studies

Ana M. Ferreira[1,3], Laura Restelli[5], Susana S. Araújo[2,3], Francesca Dilda[5], Elvira Sales-Baptista[1,4], Fabrizio Ceciliani[5] and André M. Almeida[2,3]

[1]*Instituto de Ciências Agrárias e Ambientais Mediterrânicas (ICAAM), Universidade de Évora, Évora, Portugal; anammf@uevora.pt*

[2]*Instituto de Investigação Científica Tropical (IICT) & Centro. Interdisciplinar de Investigação em Sanidade Animal (CIISA). CVZ – Centro de Veterinária e Zootecnia, Faculdade de Medicina Veterinária, Lisboa, Portugal*

[3]*Instituto de Tecnologia Química e Biológica/Universidade Nova de Lisboa (ITQB/UNL), Oeiras, Portugal*

[4]*Departamento de Zootecnia, Universidade de Évora (UE), Portugal*

[5]*Dipartimento di Scienze Veterinarie e Sanità Pubblica, Università degli Studi di Milano, Italy*

Introduction

Among farm animals the water buffalo (*Bubalus bubalis*) is largely used for milk, meat, and fertilizer production, representing an important fraction for the economy of many countries, especially in Tropical and Mediterranean areas. Taste plays a crucial role for animal regulation of food intake, being a determinant factor for them to chose suitable and reject unsuitable foods, hence defining their ingestive behaviour. Being bitter is considered the most interesting taste modality in herbivores due to its evolution to prevent the consumption of plant toxins. It is a very important factor to consider in animal production, as it is highly related to food preferences. In ruminants, the knowledge of genes and molecular mechanisms behind bitter taste perception is scarce and most of the studies were carried out only in cattle, due to the lack of genomic data for the other species.

The main aim of this investigation was to analyse at the molecular level bitter taste in water-buffalo. More specifically, to analyse the expression of bitter taste receptors (T2R) genes identified for other ruminants in tissue samples of water buffalo. It was our aim to describe for the first time the expression of T2R genes in this species, particularly in tongue and gastrointestinal tissues. Tongue is the main organ where taste perception occurs, more precisely in gustatory papillae (Mueller *et al.*, 2005). The expression of bitter taste receptors was studied in the gastrointestinal of humans, mice and rats (Rozengurt and Sternini, 2007). No information is at the moment available for ruminants. Therefore the present investigation is the first step of an innovative and unexplored line of research, which can help to understand the molecular background of feed selectivity in ruminants, namely in terms of tolerance to the wide range of bitter tastants in their environment and the nature of their diet.

Material and methods

Tissue samples from tongue (Figure 1) containing circumvallate gustatory papillae and intestine of four different individuals one year old males of Mediterranean water-buffalo species (*Bubalus bubalis*) were obtained from a local slaughterhouse (Milan, Italy), and kept at -80 °C in appropriate medium (RNAlater, Qiagen). These samples were used to perform RNA isolation (using Trizol, LifeTechonologies, standard protocol), followed by DNase treatment (kit Fermentas) and cDNA synthesis using iScript cDNA synthesis kit (BioRad).

Qualitative expression studies on bitter taste receptors genes (T2R) were then performed by PCR on the cDNA of the samples. PCR primers were previously designed by our group for a study on T2R sequences in sheep (*Ovis aries*) and considering the published T2R sequences of cattle. Conservation degree between the two species were analysed and found to be very high, suggesting that the same primer sequences can be used to study different ruminant species (Ferreira *et al.*, 2012). Primer sequences are represented on Table 1. Eight T2R were analysed. A sample of genomic DNA from sheep was used as positive control for the PCR

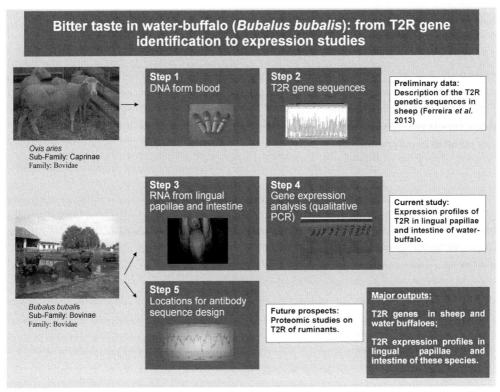

Figure 1. Overview of the project, including preliminary work, presented study and future prospects.

Table 1. Sequences of the forward and reverse primers used for PCR and sequencing of the eight T2R genes.

Genes (*Ovis aries*)	Acession number	Forward primer sequences 5'-3'
T2R16	JQ993463.1	(TGTCATAGTGCTGGGCAGAG,GGCTCTCTGCTGGTTTCTTG)
T2R10B	JQ993464.1	(CAGTGGAAGGCCTCCTAATTT,GGGTCACATCAACAGGGTTC)
T2R12	JQ993465.1	(TGGAGAGAACACTGAACAATATACTTA,AATTTCCTTTTCCTCTGGATG)
T2R3	JQ993466.1	(AGCAATTTGGGGTTTCTGGT,CCTCTGGCTCAAATGGAGAG)
T2R4	JQ993467.1	(TTTTTCTTCTATCGTTGTCTCTGAAA,TGCTGCTGGTCTGTATGCTC)
T2R67 (42)	JQ993468.1	(ATGCCATCTGGAATTGAAAA,TGCTTCTCTTGGGGTCTTTG)
T2R13	JQ993469.1	(TGGCAGATTCTTTGGAAAACA,ACAGGCCTCAGCCTCTTTTA)
T2R5 (39)	JQ993470.1	(CCAAACCTGCAGTTCCTCAG,CCTTGGGCCAGAGTGTGC)

Genes (*Ovis aries*)	Acession number	Reverse primer sequences 5'-3'
T2R16	JQ993463.1	(TCTCAAGCATGGTGCTGTTC,TCTTTTCAGTTTGGCACTGCT)
T2R10B	JQ993464.1	(GCCAGGTTGAGCTTCTGTTC,TTCTCTTTTCCCCAGCACTT)
T2R12	JQ993465.1	(AAGAGAAGACAGCCCCCAGT,CATCACTTCAGGCTTATTTTTGG)
T2R3	JQ993466.1	(CAGCAGGGTACTCAACAGCA,TGGATAAACAGATCCCTTGGA)
T2R4	JQ993467.1	(CCAAAGACAGGAGTCCCTCA,TGCTTTTGTTTTCAGTTTAGGATG)
T2R67 (42)	JQ993468.1	(TGATTCAGAATTCAAAAACAGTAAGG,TTTCAGGCGGCAGTTAAGAT)
T2R13	JQ993469.1	(TTTGGAAAAGGTTTCAGTGTCA,CACAGCACCAAAAGTGAAGC)
T2R5 (39)	JQ993470.1	(GCCTCCGAGAAGGACTTTTT,TTTAAAATAAAGGTGGACTTGAGC)

reactions. Standard conditions of PCR were used and optimized for these primers, as described (Ferreira *et al.*, 2012). Qualitative PCR was also performed on one of the T2R receptors using cDNA samples available from several parts of the digestive system of cattle (reticulum, rumen, abomasum, omasum, small intestine and caeco).

Another purpose of this study was to predict the 3D structure of the T2R receptors in order to design specific antibodies as a follow-up study. This will improve current proteomic studies on T2R research, particularly for ruminant species.

Results and discussion

The protocol of tissue collection proved to be adequate as the yield of RNA recovery from these tissues was of high-quality and appropriate for further expression studies. RNA samples were obtained from papillae and intestine of the four individuals of *Bubalus bubalis* species.

From the eight T2R genes analysed by qualitative PCR on the cDNA, the expression of six of them was detected on tongue samples whilst not found in the intestine (Figure 1).

These results confirm the conservation of T2R gene sequences among species, as the primers designed for sheep T2R work on the buffalo samples. Expression of several receptors is found in tongue as expected from the existing data on several mammals (Dong *et al.*, 2009). The results of the intestine samples suggest the lack of expression of T2R in water buffaloes. This finding is new, but it should also not be ruled out the possibility that the primers are not specific for the identification of T2R in water buffaloes samples. The expression of T2R in gastrointestinal samples has only been studied in humans, mice and rats, where T2R expression was demonstrated. Our results can point to different needs of bitter detection and prevention in ruminants outside the oral context. Ruminants have very different anatomy and physiology of digestive system and the complexity of their digestive organs can result in a no-need for additional mechanisms to avoid plant's bitter compounds consumption or toxin removal.

To further explore this hypothesis, we analysed the expression of one of the T2R receptors in cDNA samples of several parts of the digestive system of cattle (reticulum, rumen, abomasum, omasum, small intestine and caeco) and we also did not obtain amplification for all these tissues (results not shown). Gene expression results have to be confirmed and validated by the identification of the protein the tissues. Since no species-specific antibody is still available, an *in silico* investigation focused on 3D structure of T2R was carried out, in order to identify sequence domains suitable to design antigen for specific antibody production.

As future prospects, quantitative PCR should be performed to confirm and quantify the expression results, in order to draw a more refined picture of the expression levels of the various T2R genes. Nevertheless, these data serves as preliminary results for continuing this line of research. Also extending the study to different parts of the digestive system of ruminants and more T2R receptors appear as an appealing follow-up of this study, as well as pursuing the establishment of T2R antibodies specific for ruminants that allow new proteomics-based studies of bitter taste for these animal species.

Acknowledgements

This work was performed in the frame of a COST Short-Term Scientific Mission (Reference code: COST-STSM-ECOST-STSM-FA1002-160712-019235).

Authors acknowledge also financial support from *Fundação para a Ciência e a Tecnologia* (Lisboa, Portugal) in the form of the grants SFRH/BPD/69655/2010 (AM Ferreira) and Research Contract by the Ciência 2007 and 2008 programs (respectively AM Almeida and SS Araújo).

References

Dong, D., Jones, G. and Zhang, S., 2009. Dynamic evolution of bitter taste receptor genes in vertebrates. BMC Evol Biol 9: 12.

Ferreira, A.M., Araujo, S.S., Sales-Baptista, E. and Almeida, A.M., 2012. Identification of novel genes for bitter taste receptors in sheep (*Ovis aries*). Animal: 1-8.

Mueller, K.L., Hoon, M.A., Erlenbach, I., Chandrashekar, J., Zuker, C.S. and Ryba, N.J., 2005. The receptors and coding logic for bitter taste. Nature 434: 225-229.

Rozengurt, E. and Sternini, C., 2007. Taste receptor signaling in the mammalian gut. Curr Opin Pharmacol 7: 557-562.

Seasonal weight loss in dairy goats from the Canary Islands: towards an integrated omics approach?

Joana R. Lérias[1], Lorenzo E. Hernández-Castellano[2], Antonio Morales-delaNuez[2,3], Susana S. Araújo[1], Noemí Castro[2], Anastasio Argüello[2], Juan Capote[4] and André M. Almeida[1,5]

[1]Instituto de Investigação Científica Tropical, Centro de Veterinária e Zootecnia, Faculdade de Medicina Veterinária, Av. Univ. Técnica, 1300-477 Lisboa, Portugal and IBET, Av. República, 2780-157 Oeiras, Portugal; joanalerias@gmail.com
[2]Facultad de Veterinaria, Universidad de Las Palmas de Gran Canaria, Arucas, Spain
[3]Universidad Técnica de Manabí, Manabí, Ecuador
[4]ICIA – Instituto Canario de Investigaciones Agrarias, Valle Guerra, Tenerife, Spain
[5]ITQB – Instituto de Tecnologia Química e Biológica, Oeiras, Portugal

Introduction

Seasonal weight loss (SWL) is very important for animal production in the Tropics and the Mediterranean. These environments are characterized by two seasons: rainy with abundant pastures, and dry, with poor pastures where animals may lose up to 30% of their live weight with severe consequences for production performances. Therefore, SWL is the major constraint to animal production in tropical regions, which emphasizes the need to improve our knowledge in SWL and possible solutions to overcome it (Cardoso and Almeida, 2013). Presently, about half of world's sheep and goats live in the tropical regions and, additionally, the dairy industry importance has increased significantly. Therefore, it is a main concern to determine solutions for SWL, particularly as the costs associated to supplementation are particularly high (Cardoso and Almeida, 2013). The identification of markers of tolerance to SWL susceptible of being used in animal selection is considered to be of major significance to increase stock productivity (Eckersall et al., 2012). In this work, we used two different goat breeds from the Canary Islands (Spain) with contrasting levels of adaptation to SWL: Majorera (adapted to nutritional stress) and the Palmera breed (susceptible to subnutrition). Both breeds have common ancestors (Amills et al., 2004), that may serve as a model for other goat breeds. The ultimate objective of this work is to identify biomarkers of tolerance to weight loss, using proteomics and transcriptomics studies that may provide information regarding different goats' breeds' tolerance to SWL.

Material and methods

The study was conducted during May/June 2012 at the experimental farm of the Faculty of Veterinary Medicine of the ULPGC (Arucas, Gran Canaria, Spain) with 10 Majorera and 10 Palmera adult dairy goats (3 lactations with kidding in late February) obtained from the experimental flock of the Pico Research Station (Valle Guerra, Tenerife, Spain). Goats were divided in four sets, two for each breed: an underfed group in which animals were fed on wheat straw *ad libitum* (restricted diet, so a 15-20% reduction of their initial body weight would be attained by the end of the experimental period), and a control group in which animals were fed *ad libitum* on maize, soy 44 (crude protein 44%), dehydrated lucerne, dehydrated beetroot, lucerne hay and a vitamin-mineral supplement, in accordance with the guidelines issued by INRA as previously described (Martinez-de la Puente *et al.*, 2011). Goats were weighted on days 0, 9, 13, 14, 15, 16, 17, 20, 21, 22 and 23 of the trial. Animals were milked once-daily in a milking parlor equipped with recording jars (4 liters ± 25 per ml) and the milk yield was recorded on the same days of weight record and also on day 10. Statistical analysis was conducted as described by (Almeida *et al.*, 2013). This experiment allowed us to ascertain the weight and production losses, in terms of milk output, in the two goat breeds, creating mammary gland samples that will be used in proteomics and transcriptomics studies.

Results and discussion

The evolution of relative live weights is presented in Figure 1A. The animals from both control groups had an increase of weights of approximately 7% and 4% for *Majorera* and *Palmera*,

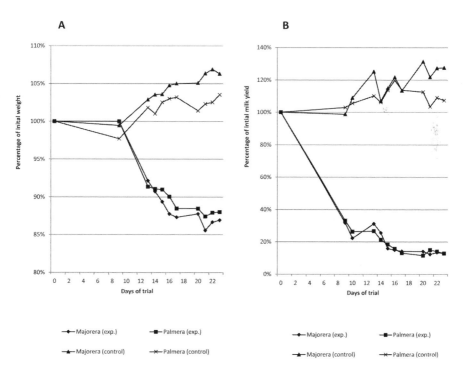

Figure 1. Relative evolution of live weights (A) and milk production (B) for the Majorera *and* Palmera *goats in the trial. Control groups have significant differences regarding to underfed groups from day 9 to 23 (P<0.05).*

respectively. In relation to the experimental groups, both had a reduction of their live weights of approximately 14% and 13% for *Majorera* and *Palmera*, respectively. The evolution of relative milk yields is presented in Figure 1B. Both control groups increased their milk yields during the studied period in 31% and 20% (*Majorera* and *Palmera*, respectively). Concerning the experimental groups both had their values reduced for a total of approximately 88% of their initial milk yield. In relation to the absolute live weights and milk yields for all groups, their values are presented in Table 1 and 2, respectively. In general, *Majorera* control group had greater increments of live weights and milk yields than *Palmera* control group. In relation to the underfed groups, their live weights reductions were slightly greater for the *Majorera* breed. However, considering milk production, both experimental groups had similar reductions of those values along the trial.

Future perspectives

The results herein presented are a preliminary phase of a larger project focused in determining biomarkers associated to SWL tolerance using several omics tools. SWL is the major constraint for the livestock producers in tropical climates, producing significant economic losses. Additionally, associating these with the increased significance that small ruminants have achieved in animal production in tropical and Mediterranean climates, it can be demonstrated the importance to improve our knowledge on the physiology of SWL and, therefore, to be able to determine new mechanisms to overcome it. In the near future, mammary gland biopsies from the animals of the trial herein described will be studied to compare their transcriptomes

Table 1. Absolute live weights of Majorera and Palmera control and experimental groups.[1,2]

| | Days of trial | | | | | | | | | | |
	0	9	13	14	15	16	17	20	21	22	23
Majorera C	45.5[a]	45.2[a]	46,7[a]	47.0[a]	47.0[ab]	47.6[ab]	47.7[ab]	47.6[ab]	48.3[ab]	48.5[a]	48.2[ab]
	±7.74	±7.76	±7.46	±7.67	±7.85	±7.77	±8.07	±7.20	±7.72	±7.50	±7.51
Palmera C	32.8[bc]	32.1[bc]	33.4[bc]	33.2[bc]	33.6[c]	33,7[c]	33.9[c]	33.2[c]	33.5[c]	33.6[c]	33.9[c]
	±4.91	±4.90	±5.26	±5.06	±4.86	±4.74	±5.25	±4.76	±4.74	±4.99	±5.00
Majorera S	50.6[a]	46.6[a]	45.9[a]	45.3[a]	44.4[a]	44.2[a]	44.4[a]	43.3[a]	43.9[a]	44.0[a]	44.1[a]
	±3.64	±3.49	±3.36	±3.66	±3.52	±3.17	±3.46	±3.28	±3.65	±3.46	±3.42
Palmera S	40.6[ac]	37.1[ac]	36.9[ac]	36.9[ac]	36.5[ac]	35.9[ac]	35.9[ac]	35.5[ac]	35.7[ac]	35.7[ac]	35.4[ac]
	±2.05	±2.21	±2.06	±2.33	±2.20	±1.83	±2.02	±1.75	±1.77	±1.53	±1.45

[1] C = Control group; S = Experimental group.

[2] Values are means ± standard deviation; columns with different superscripts ([a,b,c]) indicate significant differences ($P<0.05$) between experimental groups.

Table 2. Absolute daily milk yields of Majorera *and* Palmera *control and experimental groups.*

| | Days of trial | | | | | | | | | | | |
	0	9	10	13	14	15	16	17	20	21	22	23
Majorera C	1.60[a]	1.56[ac]	1.69[a]	1.93[a]	1.66[ac]	1.78[a]	1.90[a]	1.78[a]	2.05[a]	1.90[a]	1.99[a]	1.99[a]
	±0.47	±0.42	±0.32	±0.40	±0.40	±0.43	±0.42	±0.41	±0.49	±0.44	±0.45	±0.48
Palmera C	1.03[a]	1.08[bc]	1.11[ac]	1.15[b]	1.14[bc]	1.20[a]	1.27[a]	1.20[a]	1.21[a]	1.11[b]	1.16[b]	1.15[b]
	±0.51	±0.57	±0.62	±0.60	±0.66	±0.68	±0.73	±0.68	±0.74	±0.65	±0.64	±0.67
Majorera S	1.68[a]	0.52[b]	0.37[b]	0.52[b]	0.42[b]	0.26[b]	0.25[b]	0.24[b]	0.23[b]	0.21[c]	0.23[c]	0.22[c]
	±0.25	±0.06	±0.04	±0.08	±0.03	±0.05	±0.07	±0.11	±0.07	±0.03	±0.04	±0.05
Palmera S	1.33[a]	0.44[b]	0.35[bc]	0.36[b]	0.29[bc]	0.25[b]	0.21[b]	0.18[b]	0.15[b]	0.20[c]	0.19[c]	0.17[c]
	±0.19	±0.05	±0.09	±0.16	±0.14	±0.12	±0.11	±0.06	±0.06	±0.07	±0.06	±0.01

[1] C = Control group; S = Experimental group.

[2] Values are means ± standard deviation; columns with different superscripts ([a,b,c]) indicate significant differences (*P*<0.05) between experimental groups.

and proteomes using 'omics' tools. We will use a molecular approach to understand the mechanisms underlying an effective lactation during SWL, aiming to associate the obtained molecular data with physiological and productive performance. The transcriptomics tasks will involve the use of microarrays and quantitative PCR. In relation to proteomics, we will use two-dimensional electrophoresis/DIGE and protein identification through mass spectrometry. The results of these studies will provide novel information on the identification of possible candidate genes and proteins that are associated to a successful lactation during adverse conditions as SWL. In summary, this project will allow the identification of candidate genes and proteins associated with SWL tolerance, which will be used to improve small ruminant production in tropical and Mediterranean climates.

Acknowledgements

This research was supported by the research project PTDC/CVT/116499/2010 – Lactation and milk production in goat (*Capra hircus*): identifying molecular markers underlying adaptation to seasonal weight loss from Fundação para a Ciência e Tecnologia (FCT, Lisbon, Portugal). Authors AM Almeida and SS Araújo are funded respectively by the Ciência 2007 and Ciência 2008 Programs by the FCT. Authors acknowledge the staff of the *Pico* Research Station and the ULPGC Experimental farm.

References

Almeida, A.M., Kilminster, T., Scanlon, T., Araujo, S.S., Milton, J., Oldham, C. and Greeff, J.C., 2013. Assessing carcass and meat characteristics of Damara, Dorper and Australian Merino lambs under restricted feeding. Trop Anim Health Prod.

Amills, M., Capote, J., Tomas, A., Kelly, L., Obexer-Ruff, G., Angiolillo, A. and Sanchez, A., 2004. Strong phylogeographic relationships among three goat breeds from the Canary Islands. J Dairy Res 71: 257-262.

Cardoso, L.A. and Almeida, A.M., 2013. Seasonal weight loss – an assessment of losses and implications on animal welfare and production in the tropics: Southern Africa and Western Australia as case studies. Enhancing Animal welfare and farmer income through strategic animal feeding. Makkar, H.P.S. (ed) FAO Animal Production and Heath paper No. 175. Food and Agricultural Organization of the United Nations (FAO), Rome.

Eckersall, P.D., de Almeida, A.M. and Miller, I., 2012. Proteomics, a new tool for farm animal science. J Proteomics 75: 4187-4189.

Martinez-de la Puente, J., Moreno-Indias, I., Morales-Delanuez, A., Ruiz-Diaz, M.D., Hernandez-Castellano, L.E., Castro, N. and Arguello, A., 2011. Effects of feeding management and time of day on the occurrence of self-suckling in dairy goats. Vet Rec 168: 378.

Analysis of endogenous and recombinant bovine somatotropin in serum

Marco H. Blokland[1], Klaas L. Wubs[1], Merel A. Nessen[1], Saskia S. Sterk[1], and Michel W.F. Nielen[1,2]
[1]*RIKILT Wageningen UR, European Union Reference Laboratory for Residues, Wageningen, the Netherlands; marco.blokland@wur.nl*
[2]*Laboratory of Organic Chemistry, Wageningen UR, Wageningen, the Netherlands*

Objectives

Within the EU the use of growth hormones is banned. However, in several countries outside the EU the use of the growth hormone somatotropin, a 191 amino acid protein, for cattle is allowed. In the United States recombinant bovine somatotropin (rbST) is allowed to increase milk production in cows, while recombinant porcine somatotropin (rpST) is registered in Australia as a growth promoter for pigs (Blokland *et al.*, 2003). Given the good/positive effect on, amongst others, animal growth and milk production and the scale of production of these proteins (by recombinant technology) it is expected that somatotropin is used illegally as growth promoter in Europe. Consequently effective analytical methods are needed for residue control.

This poster describes the development of a method for the detection and identification of endogenous and recombinant somatotropin in biological materials. The difference in the N-terminal amino acid of the sequence of the endogenous and recombinant protein (alanine for bST, methionine for rbST (Pinel *et al.*, 2004)) is used to discriminate between the two proteins. After total serum digestion, these unique peptides for rbST and bST are isolated by solid phase extraction (SPE) and detected by liquid chromatography – triple quad mass spectrometry (LC-QqQ-MS/MS). For quality control and quantification isotope labelled peptides are used. Typical pitfalls related to the analytical methodology of somatotropin analysis are presented and discussed.

Material and methods

All chemicals and reagents were of highest purity and quality. The use of special 'low-bind' tubes for protein samples such as Eppendorf LoBind tubes is highly recommended.

500 µl serum was digested with trypsin at pH=8.2 under sequential acetonitrile addition based on a protocol published before (Arsene *et al.*, 2008). After stabilizing the pH of the serum to pH=8.2 by the addition of 100 µl of 0.1 M tris-buffer, 50 µl of a solution of 20 µg/µl trypsin in 50 mM acetic acid was added to start the digestion. To slow-down the autolysis of trypsin 0.02 mmol calcium chloride was added to the digestion solvent. The solution

was incubated at 37 °C under continuous shaking. Isotope labelled peptides, similar to the peptides of endogenous and recombinant somatotropin (MFPAMSLSGLFANAVLR for rbST and AFPAMSLSGLFANAVLR for bST), were added as internal standards. After 10, 30, 90, 150, and 210 min additional trypsin solution (50 µl) was added to the reaction mixture. In parallel, every 30 min (starting at t=30 min) 100 µl acetonitrile was added, resulting in a final concentration of 58% acetonitrile after 330 min. The reaction was stopped after 24 hours.

Clean-up of the digest was necessary to extract and concentrate the peptides specific for bST and rbST from the digestion mixture. Primary clean-up was accomplished by HPLC-fractionation on a C18 column (Zorbax 300 Extend-C18, 2.1×150 mm, 3.5 µm) using a gradient with water, acetonitrile and 0.1% formic acid. The fraction containing the peptides of (r)bST was collected. This fraction was concentrated on a Varian Bond Elut Plexa SPE column (60 mg/3 ml). After conditioning the column with 2×1 ml methanol and 1 ml 1% formic acid in water, the sample, which was diluted to 9 ml final volume with 5% aqueous formic acid, was applied onto the column. The column was washed with 1 ml 5% methanol, containing 0.2% formic acid and finally eluted in a tube containing 100 µl dimethyl sulfoxide (DMSO) with sequentially 1 ml and 0.5 ml water/methanol/formic acid (45/50/5).

The eluate from the SPE was evaporated to approx. 100 µl on a Zymark TurboVap at 55 °C under 10 psi N_2. After cooling to room temperature, 100 µl 5% formic acid was added to reconstitute the dried extract. The sample was mixed and, after ultrasonic treatment, transferred to an injection vial.

The final extract is analysed by LC-MS/MS in multiple reaction monitoring (MRM)-mode. The chromatographic separation was performed on a Waters Acquity UPLC with a BEH C_{18} column (2.1×100 mm, 1.7 µm), using a gradient containing water, acetonitrile and 0.1% formic acid. The Waters XEVO-TQS mass spectrometer was operated in the ESI-MS/MS positive mode, measuring two transitions for each peptide of interest and one transition for the internal standards.

Results and discussion

In the literature several screening and confirmation methods are described to detect endogenous and recombinant somatotropin in serum (Le Breton et al., 2008; Smits et al., 2012), and more recently, in milk (Le Breton et al., 2010; Ludwig et al., 2012). The confirmation methods are based on specific isolation of somatotropin from the matrix, followed by a tryptic digestion step and the isolation of the N-terminal peptide. Since there is no isotope labelled somatotropin available it is difficult to check and correct for losses during the clean-up. As somatotropin is present at very low concentration (1-10 ng/ml) in serum samples, it is difficult to isolate the whole protein using a simple and fast procedure. Moreover, the described clean-up procedures result in a very low recovery of the protein, making the detection of the peptides of interest even more difficult.

In this study we investigated an alternative approach to detect somatotropin in a biological matrix. Instead of first isolating the whole protein followed by a tryptic digestion step, we started with a total-serum digestion using trypsin, followed by an extraction procedure based upon a HPLC fractionation and a solid phase clean-up concentration. Obtained peptide mixtures were analysed with LC-MS/MS using MRM mode, targeting the N-terminal peptides specific for the endogenous and recombinant somatotropin.

In a previous study it was found mass spectrometric analysis of the peptides of interest was not successful due to the poor solubility and extreme adsorption of the peptides of interest to glassware, plastics or other materials used. Here, different solvents were tested to reconstitute the peptides and to keep the peptides in solution. DMSO was found to be most effective in keeping the peptides in solution and for reconstitution of the peptides after evaporation, without interfering with mass spectrometric analysis.

To test the developed method, serum samples were spiked with endogenous and recombinant somatotropin at a level of 50 ng/ml. Then, as the internal standard, isotope labelled peptides were added to the serum. Samples were digested with trypsin and the obtained peptides mixtures were separated on an HPLC-system as described in the material and method section. The fractions of interest were collected, concentrated and analysed by targeted LC-MS/MS. The selected transitions for the N-terminal peptides were found to be selective for both bST and rbST in serum as is shown in Figure 1. The use of isotope labelled peptides as internal standards makes it possible to perform robust quality control during the whole clean-up

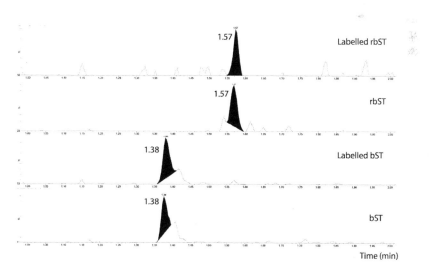

Figure 1. MRM chromatograms of N-terminal peptides of bST and rbST and the isotope labelled corresponding peptides. Example of the targeted LC-MS/MS analysis of a fully processed sample containing bST and rbST spiked at 50 ng/ml. For each peptide one specific transition is shown.

process. This allows specific detection of natural occurring bST and the recombinant rbST in serum samples.

To conclude, a method has been developed that can be used to isolate and analyse somatotropin in spiked samples. It is possible to detect bovine somatotropin (bST) and recombinant somatotropin (rbST) in fully processed samples at a level of 50 ng/ml. This level is still above relevant concentration levels in serum, which are typically in the order of 1-10 ng/ml. It is expected that through the use of nano-UPLC-MS/MS these levels will be in reach of the described method.

References

Arsene, C.G., Ohlendorf, R., Burkitt, W., Pritchard, C., Henrion, A., O'Connor, G., Bunk, D.M. and Guttler, B., 2008. Protein quantification by isotope dilution mass spectrometry of proteolytic fragments: Cleavage rate and accuracy. Anal Chem 80: 4154-4160.

Blokland, M.H., Sterk, S.S., van Ginkel, L.A., Stephany, R.W. and Heck, A.J.R., 2003. Analysis for endogenous and recombinant porcine somatotropine in serum. Anal Chim Acta 483: 201-206.

Le Breton, M.H., Beck-Henzelin, A., Richoz-Payot, J., Rochereau-Roulet, S., Pinel, G., Delatour, T. and Le Bizec, B., 2010. Detection of recombinant bovine somatotropin in milk and effect of industrial processes on its stability. Anal Chim Acta 672: 45-49.

Le Breton, M.H., Rochereau-Roulet, S., Pinel, G., Bailly-Chouriberry, L., Rychen, G., Jurjanz, S., Goldmann, T. and Le Bizec, B., 2008. Direct determination of recombinant bovine somatotropin in plasma from a treated goat by liquid chromatography/high-resolution mass spectrometry. Rapid Commun Mass Spectrom 22: 3130-3136.

Ludwig, S.K.J., Smits, N.G.E., Bremer, M.G.E.G. and Nielen, M.W.F., 2012. Monitoring milk for antibodies against recombinant bovine somatotropin using a microsphere immunoassay-based biomarker approach. Food Control 26: 68-72.

Pinel, G., Andre, F. and Le Bizec, B., 2004. Discrimination of recombinant and pituitary-derived bovine and porcine growth hormones by peptide mass mapping. J Agric Food Chem 52: 407-414.

Smits, N.G.E., Bremer, M.G.E.G., Ludwig, S.K.J. and Nielen, M.W.F., 2012. Development of a flow cytometric immunoassay for recombinant bovine somatotropin-induced antibodies in serum of dairy cows. Drug Test Anal 4: 362-367.

Investigation of salivary acute phase proteins in calves

Mizanur Rahman[1], Ute Müller[1], Peter Heimberg[2], Fabrizio Ceciliani[3] and Helga Sauerwein[1]
[1]Institute of Animal Science, Physiology & Hygiene Unit, University of Bonn, Germany;
l-mrah@uni-bonn.de
[2]Landwirtschaftskammer Nordrhein-Westfalen, Referat 34, Tiergesundheitsdienste, Münster,
Germany
[3]Department of Veterinary Science and Public Health, University of Milan, Italy

Introduction

Saliva is a colorless viscous liquid secreted from salivary glands that plays a major role in oral and general health. Saliva shares several compositional similarities with other body fluids e.g. blood and urine and is therefore interesting for biological assays (Lac *et al.*, 1993). When compared to other biological samples as analytical matrix, saliva is advantageous for practical purposes due to its simple and minimally stressful, non-invasive collection. In addition to being more straightforward and economical to obtain than blood, saliva has the further advantage of being easier to handle for diagnostic purposes because it does not clot (Wong, 2006). The use of saliva for analytical purposes in mostly focused on clinical investigations (i.e. in the field of endocrinology, neuroendocrinology) and in physiological research (i.e. in sport and exercise science). Recent developments comprise proteomic and metabolomic analyses of saliva for identifying markers of disease (Lamy, 2012), however, the most frequent application of saliva analysis *in praxi* is probably the assessment of cortisol as a biomarker of stress. In veterinary medicine and in animal science, salivary cortisol is considered particularly important to assess welfare under different conditions of keeping and handling. However, besides cortisol, the use of saliva in veterinary science is still quite limited. With regard to acute phase proteins (APP) in animal saliva, the available literature is largely limited pigs and dogs. The APPs are blood proteins that can be monitored to assess the innate immune system's systemic response to infection, inflammation, and trauma or stress (Ceron *et al.*, 2005; Murata *et al.*, 2004). Serum amyloid A (SAA), C-reactive protein and haptoglobin (Hp) have been identfied and quantified in porcine saliva and were demonstrated to be well correlated with the circulating concentrations (Gutierrez *et al.*, 2009; Hiss *et al.*, 2003; Hiss, 2009; Soler *et al.*, 2012). To the best of our knowledge, there are no data available about the presence of APP in ruminant saliva. Alpha 1 acid glycoprotein (AGP), lipopolysacccharide binding protein (LBP), Hp and SAA have been reported to be expressed in bovine salivary glands (Lecchi *et al.*, 2009; Lecchi *et al.*, 2012; Rahman *et al.*, 2010), thus their presence in saliva is probable and might be due to local expression and also to transfer from blood. In order to investigate the presence of four major APP, i.e. AGP, Hp, LBP and SAA in bovine saliva, we used samples collected from calves before and after dehorning. Dehorning of calves is a routine procedure carried out under local anesthesia with the purpose to facilitate management in later life. Dehorning is considered as stressful but the lesions are probably only mild since an acute phase reaction is reportedly not occurring (Doherty *et al.*, 2007). Using quantitative assays for Hp as well as qualitative Western Blots (WB) for AGP, LBP and SAA, we aimed to evaluate whether these APP are detectable in saliva.

Materials and methods

Saliva was collected from 34 calves just before dehorning (0 h), 1 h and 24 h after dehorning; a blood sample was also drawn 24 h after dehorning. Hp in saliva and in blood serum was measured by ELISA (Hiss *et al.*, 2003).

Total protein (TP) was quantified by the Bradford method. An appropriate amount of TP from saliva or serum samples depending upon the specific APP (for details see legend of Figure 1) was separated by sodium dodecyl sulphate polyacrylamide gel electrophoresis (SDS-PAGE) and

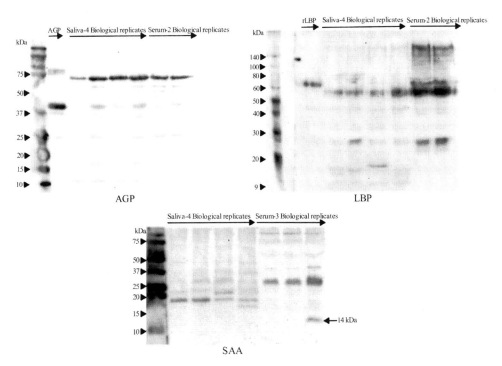

Figure 1. Western blot analyses of AGP, SAA and LBP in calf saliva and serum. The APPs were identified after immunostaining using the antibodies listed in Table 1 and detection by ECL. The amounts of total saliva protein per lane were: 5 μg (AGP), 10 μg (LBP) and 20 μg (SAA), respectively. The serum volumes (μl) used were 7.5 (AGP), 5 (LBP) and 19 (SAA), respectively. The saliva and serum samples applied were randomly selected from the dehorning trial and pools were generated to ensure sufficient amounts of each sample for all WBs.

blotted onto polyvinyl difluoride (PVDF) membranes. The membranes were immunolabelled for the presence of APPs using specific antibodies (Table 1) and immunoreactive bands were visualized by enhanced chemiluminescence (ECL). Recombinant human (rh) LBP (Biometec Ltd, Greifswald, Germany), purified bovine AGP and bovine serum were used as positive controls.

Results

Comparing the salivary Hp concentrations at the different times of sampling relative to dehorning, increased Hp values were recorded 1 h (23.5 ± 35.5 µg/ml) after dehorning as compared to 0 h (9.38 ± 10.3 µg/ml) ($P<0.05$, mixed model, SPSS), however, the increase was mild, i.e. 2.5 fold as compared to 0 h. At 24 h, baseline values were reached again. The Hp concentrations in the 24 h saliva and serum samples were correlated ($r=0.647$, $P<0.001$).

The presence of AGP, LBP and SAA in bovine saliva was confirmed by WB analysis. Exemplary blots are shown in Figure 1. For AGP, a protein band around 66 kDa was detectable both in saliva and serum. In addition, there were a faint 45 kDa band and other low molecular weight (MW) bands (Figure 1). For SAA, the bands obtained differed between saliva and serum: in saliva, a 19 kDa band was observed, whereas in serum a 26 kDa reactive band was detected. Comparing the different pools, more intense salivary bands were observed in samples from 0 h than in the 24 h sample (Figure 1).

For LBP, bands were detected mainly at 55 kDa. Additional bands of 25 and 20 kDa were present both in saliva and in serum (Figure 1).

Discussion and conclusion

Saliva from calves can easily be collected with minimal stress. In pigs saliva APP measurements have been demonstrated to be useful for monitoring infectious diseases (Gutierrez *et al.*, 2012; Soler *et al.*, 2012). APP can play an imperative role to monitor the systemic response to infections, trauma, stress or other noxa. The presence of AGP at different MW band size can be explained by different glycoforms of this APP. Differences in SAA band pattern may be due to different glycan moiety. The low MW band of LBP might be associated with same

Table 1. APP antibodies used for Western blot analyses

Primary antibody	Dilution	Incubation time (h)	Gel type
Rabbit anti-bovine AGP	1:2,000	0.45	12% SDS-PAGE
Mouse anti-human LBP	1:600	16 (Overnight)	12% SDS-PAGE
Rabbit anti-bovine SAA	1:25,000	1	15% Tris-tricine-SDS

protein (Rahman *et al.*, 2010). Based on the qualitative proof of the presence of SAA, AGP and LBP, future research should provide quantitative data to characterize the informational value of salivary APP.

Acknowledgement

The award of research fellowship from Alexander von Humboldt Foundation, Germany to Md. Mizanur Rahman is gratefully acknowledged.

References

Ceron, J.J., Eckersall, P.D. and Martynez-Subiela, S., 2005. Acute phase proteins in dogs and cats: current knowledge and future perspectives. Vet Clin Pathol 34: 85-99.

Doherty, T.J., Kattesh, H.G., Adcock, R.J., Welborn, M.G., Saxton, A.M., Morrow, J.L. and Dailey, J.W., 2007. Effects of a concentrated lidocaine solution on the acute phase stress response to dehorning in dairy calves. J Dairy Sci 90: 4232-4239.

Gutierrez, A.M., Ceron, J.J., Fuentes, P., Montes, A. and Martinez-Subiela, S., 2012. Longitudinal analysis of acute-phase proteins in saliva in pig farms with different health status. Animal 6: 321-326.

Gutierrez, A.M., Martinez-Subiela, S. and Ceron, J.J., 2009. Evaluation of an immunoassay for determination of haptoglobin concentration in various biological specimens from swine. Am J Vet Res 70: 691-696.

Hiss, S., Knura-Deszczka, S., Regula, G., Hennies, M., Gymnich, S., Petersen, B. and Sauerwein, H., 2003. Development of an enzyme immuno assay for the determination of porcine haptoglobin in various body fluids: testing the significance of meat juice measurements for quality monitoring programs. Vet Immunol Immunopathol 96: 73-82.

Hiss, S., Weinkauf, C, Hachenberg, S, Sauerwein, H., 2009. Short communication: Relationship between metabolic status and the milk concentrations of haptoglobin and lactoferrin in dairy cows during early lactation. J Dairy Sci 92.

Lac, G., Lac, N. and Robert, A., 1993. Steroid assays in saliva: a method to detect plasmatic contaminations. Arch Int Physiol Biochim Biophys 101: 257-262.

Lamy, E., and Mau, M., 2012. Saliva proteomics as an emerging, non-invasive tool to study livestock physiology, nutrition and diseases. J Proteomics 75: 4251-4258.

Lecchi, C., Avallone, G., Giurovich, M., Roccabianca, P. and Ceciliani, F., 2009. Extra hepatic expression of the acute phase protein alpha 1-acid glycoprotein in normal bovine tissues. Vet J 180: 256-258.

Lecchi, C., Dilda, F., Sartorelli, P. and Ceciliani, F., 2012. Widespread expression of SAA and Hp RNA in bovine tissues after evaluation of suitable reference genes. Vet Immunol Immunopathol 145: 556-562.

Murata, H., Shimada, N. and Yoshioka, M., 2004. Current research on acute phase proteins in veterinary diagnosis: an overview. Vet J 168: 28-40.

Rahman, M.M., Lecchi, C., Avallone, G., Roccabianca, P., Sartorelli, P. and Ceciliani, F., 2010. Lipopolysaccharide-binding protein: Local expression in bovine extrahepatic tissues. Vet Immunol Immunopathol 137: 28-35.

Soler, L., Gutierrez, A. and Ceron, J.J., 2012. Serum amyloid A measurements in saliva and serum in growing pigs affected by porcine respiratory and reproductive syndrome in field conditions. Res Vet Sci 93: 1266-1270.

Wong, D.T., 2006. Salivary diagnostics powered by nanotechnologies, proteomics and genomics. J Am Dent Assoc 137: 313-321.

Targeted proteomics as a tool for porcine acute phase proteins measurements

Anna Marco-Ramell[1], Thomas F. Dyrlund[2], Stine L. Bislev[2], Lorenzo Fraile[3,4], Kristian W. Sanggaard[2], Jan J. Enghild[2], Emøke Bendixen[2] and Anna Bassols[1]

[1]Dept. Bioquímica,Universitat Autònoma de Barcelona, Spain; marcoramell@gmail.com
[2]Dept. Molecular Biology, Aarhus University, Denmark
[3]Centre de Recerca en Sanitat Animal, Spain
[4]Dept. Producció Animal, Universitat de Lleida, Spain

Introduction

It is mandatory, after protein identification, to validate candidate markers in a larger amount of samples for biomarker discovery. The most used methods for validation of new markers are enzymatic or immunoassay-based commercial kits. However, relatively few kits are available for farm animals. Selected reaction monitoring (SRM), a targeted quantitative proteomic technique, may be used as an alternative to commercial kits for the measurement and validation of target proteins. Acute phase proteins (APPs) are widely recognized inflammation and infection biomarkers (Eckersall, 2010) and there is recent evidence that they can be also considered as welfare markers (Giannetto *et al.*, 2011). During an acute stress situation, the levels of some interleukins (IL-1β, IL-6 and TNF-α) are increased (Elenkov and Chrousos, 2002) and consequently the concentration of the APPs varies dramatically. In pigs the most important APPs include haptoglobin (Hp), C-reactive protein (CRP), the inter-α-inhibitor-heavy chain 4 (ITIH4, also called Pig major acute phase protein, or Pig-MAP), serum amyloid A (SAA) and apolipoprotein A-I (Apo A-I), but also other proteins are well known to mark the acute phase response, as albumin, transferrin (Heegaard *et al.*, 2011) and fetuin A (Brown *et al.*, 1992).

Objectives

Our aim was to design and optimize a selected reaction monitoring (SRM) based method for the measurement of six porcine APPs in serum: Hp, CRP, ITIH4, SAA, Apo A-I and fetuin A, and use this method to analyze the acute phase response of 13 sows, which were subjected to mild stress by a move from a common pen to individual boxes for insemination and gestation. This project was carried out at Aarhus University (Denmark) by means of a COST-Short Term Scientific Mission.

Methods

APPs protein sequences were obtained from UniProt and pasted into the Skyline software (MacCoss Lab Software). Then, the following peptide and transition parameters were fixed for filtering:

- Peptide settings. Digestion: Trypsin; Max missed cleavages: 1; Length: 7-20; Exclude N-term AAs: 25; Exclude Peptides Containing: Cys, Met.
- Transition settings. Precursor charges: 2; Ion charges: 1; Ion types: y.

Serum proteins were denatured, reduced, alkylated and digested with trypsin. Then, peptides were micropurified using C18 Stage Tips (Thermo Scientific, Odense, Denmark). 1.5 μg peptide of each sample was run in triplicate on an AB SCIEX QTRAP® 5500 System, using microgradients of 60 min, for method optimization, or 23 min, for serum sample analysis. Results were imported to Skyline, processed and analyzed with statistical software.

Results and discussion

SRM method optimization

Many unique peptides were obtained in Skyline for Hp, ITIH4, Apo A-I and fetuin A after filtering. We tried to avoid selecting short peptides, and peptides including Cys or Met, which are prone to modifications. Accordingly, for CRP and SAA, only a few suitable peptides were available. Porcine serum samples with high APP levels were used for method optimization. Four proteotypic peptides were selected from Hp, ITIH4 and Apo A-I, and the 3 most intense transition signals from each of these were optimized, while for fetuin A only a single peptide with high transition signals was available. CRP and SAA were excluded from the study because of the low intensity of their transitions, which were in the noise level.

Measurement of the acute phase response of 13 sows subjected to a mild stressor

Changes in protein concentrations were observed after SRM data analysis. Ratios between post-stress days and the control situation were calculated for all the proteins. SRM confirmed that Hp slightly increased after exposure to stress and returned to basal levels 48 hours later, whereas ITIH4 mean levels were not elevated (Figures A1 and A2) but a large variability between animals was observed. On the other hand, the negative APPs Apo A-I and fetuin A decreased when animals were faced with the challenging situation. Apo A-I levels were still very low at the end of the experiment and fetuin A levels returned to baseline 72 h later (Figures B1 and B2). SRM results showed also a large variability between animals. We also quantified these four APPs using immunoassay-based and colorimetric commercial kits. Porcine fetuin A was measured with two different commercial ELISAs and results were inconclusive and highly differed between these kits. A general agreement between SRM and commercial methods

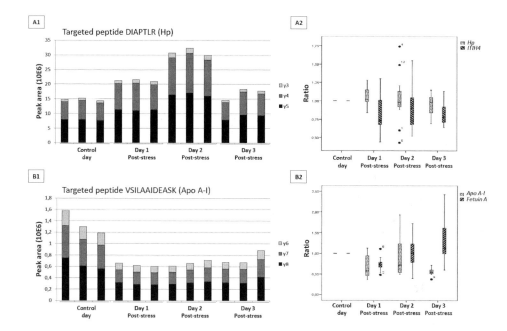

Figure 1. SRM data analysis. (A) positive APPs; A1: Bioinformatic analysis with Skyline software of one of the targeted Hp peptides (DIAPTLR); A2: Box plot of positive APPs change ratios after exposure to stress. (B) negative APPs; B1: Bioinformatic analysis with Skyline software of one/a targeted peptide from Apo A-I (VSILAAIDEASK); B2: Box plot of negative APPs change ratios after exposure to stress.

was observed: Hp (r=0.397, *P*=0.022), ITIH4 (r=0.847, *P*<0.001), Apo A-I (r=0.448, *P*=0.028) and fetuin A (ELISA1: r=0.040, *P*=0.841; ELISA 2: r=-0.269, *P*=0.137).

Conclusions

We conclude that commercial kits and SRM measures of porcine Hp, ITIH4 and Apo AI correlated well. SRM is an important alternative to the commercial kits for the quantification of porcine proteins.

References

Brown, W.M., Saunders, N.R., Mollgard, K. and Dziegielewska, K.M., 1992. Fetuin--an old friend revisited. Bioessays 14: 749-755.

Eckersall, P.D.a.B., R., 2010. Acute phase proteins: Biomarkers of infection and inflammation in veterinary medicine. Vet J 185: 23-27.

Elenkov, I.J. and Chrousos, G.P., 2002. Stress hormones, proinflammatory and antiinflammatory cytokines, and autoimmunity. Ann N Y Acad Sci 966: 290-303.

Giannetto, C., Fazio, F., Casella, S., Marafioti, S., Giudice, E. and Piccione, G., 2011. Acute phase protein response during road transportation and lairage at a slaughterhouse in feedlot beef cattle. J Vet Med Sci 73: 1531-1534.

Heegaard, P.M., Stockmarr, A., Pineiro, M., Carpintero, R., Lampreave, F., Campbell, F.M., Eckersall, P.D., Toussaint, M.J., Gruys, E. and Sorensen, N.S., 2011. Optimal combinations of acute phase proteins for detecting infectious disease in pigs. Vet Res 42: 50.

LC-MS/MS analyses of visceral and subcutaneous adipose tissue proteomes in goats

Laura Restelli[1], Marius Cosmin Codrea[2], Giovanni Savoini[1], Fabrizio Ceciliani[3] and Emoke Bendixen[2]
[1]*Department of Veterinary Science for Animal Production, Food Safety and Animal Health; Faculty of Veterinary Medicine – Università degli Studi di Milano, Milan, Italy; laura.restelli@unimi.it*
[2]*Department of Animal Science and Department of Molecular Biology and Genetics (current affiliation) – Aarhus Universitet, Viborg, Denmark*
[3]*Department of Veterinary Science and Public Health; Faculty of Veterinary Medicine – Università degli Studi di Milano, Milan, Italy*

Objectives

Adipose tissue (AT) is considered a group of highly active endocrine organs, taking part in regulating several metabolic processes, including reproduction, inflammatory response, and production and secretion of signalling molecules with important biological roles, also known as adipokines. Different deposits of visceral and subcutaneous origin differ in mRNA abundance of several adipokines in sheep (Lemor *et al.*, 2010); few studies have been carried out at a proteomic level in ruminant adipose tissue (Rajesh *et al.*, 2012) and none in goats.

The aim of this study was to perform proteomic comparisons of different types of adipose tissues, sampled at different anatomical locations. Two different levels of characterization were attempted:
1. Proteomic characterization of different subcutaneous and visceral adipose tissue deposits.
2. Comparison between proteomic maps of subcutaneous adipose tissue deposits (sternum and base of the tail), visceral adipose tissue deposits (perirenal and omental) and liver.

Materials and methods

Samples were obtained from four healthy goat-kids naturally reared by their mothers. They were slaughtered at one month age, and different adipose tissues were sampled from subcutaneous depots (sternum and base of the tail), visceral depots (perirenal and omental) as well as from liver. A descriptive proteomic analysis was accomplished through a shotgun approach at the Department of Animal Science at Aarhus University, Denmark.

Proteins were obtained after precipitation with six volumes of ice-cold acetone, using the same protocol for liver and fat tissues. After tryptic digestion, peptides were fractionated using a strong cation exchange liquid chromatography; the resultant fractions were merged and further fractionated with a reverse phase liquid chromatography and analysed with a tandem mass spectrometry system (ESI-qTOF) (Danielsen *et al.*, 2011). The four biological replicates from each of the five tissues were searched separately using ProteinPilot 1.0 software; at confidence level of 95% for protein identifications. To compare the proteome profiles across the tissue types, a hierarchical clustering analysis was performed using the Ward's linkage method, an agglomerative approach based on the 'error sum of squares' when merging pairs of clusters. Data handling and analysis were performed using the statistical software package R and for the functional annotation of the identified proteins the Blast2GO tool (http://blast2go.org) (Conesa *et al.*, 2005) and the Protein Analysis Through Evolutionary Relationship (PANTHER) annotated Celera database (Thomas *et al.*, 2003) were used.

The immune related proteins were selected and a validation of the proteomic data was performed by qualitative RT-PCR in order to detect the mRNAs levels in the different tissue types.

Results and discussion

713 proteins were identified, of which 142 were expressed only in visceral deposits, 98 expressed only in subcutaneous deposits and 158 expressed only in liver.

A dendrogram grouping the five different tissues according to their protein expression profiles was generated by hierarchical clustering using the expression data of the LC-MS/MS analysis. The dendrogram showed a clear grouping in three clusters that differ in the protein expression patterns, allowing subcutaneous and visceral ATs to be distinguished by their individual proteome expression patterns, and likewise, ATs and liver were clearly distinguishable by their proteome patterns. The obtained data were searched in the Protein Analysis Through Evolutionary Relationship (PANTHER) annotated Celera database that allows clustering of the proteins into families, or functional pathways. Afterwards, proteins were grouped in four major categories, based on similarity of functions, namely: proteins involved in metabolic and proteolytic processes (G1), proteins involved in toxic response and folding (G2), proteins involved in cell adhesion, cytoskeleton and intracellular transports (G3), and proteins involved in immune and inflammatory response (G4). Proteins found to be expressed only in liver tissues were excluded from this grouping analysis. As shown in Figure 1, more than 65% of the proteins detected in AT belongs to the first group with a higher percentage of metabolic (G1) proteins found in visceral AT only compared to subcutaneous AT. 19% of the proteins are involved in toxic response and folding (G2), while only the 10% of the total AT proteome is related to cell adhesion, cytoskeleton and intracellular transports (G3). Finally 43 proteins out of 713 (6%) are related to immune and inflammatory response with a higher percentage of proteins expressed in subcutaneous tissues (7%), compared to the visceral ones (2.6%), see Figure 1a.

Since many adipokines are expressed mainly by liver, and then secreted to blood and delivered to peripheral tissues, and since the intact AT features an intense capillary network surrounding

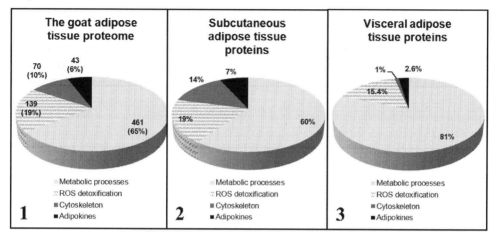

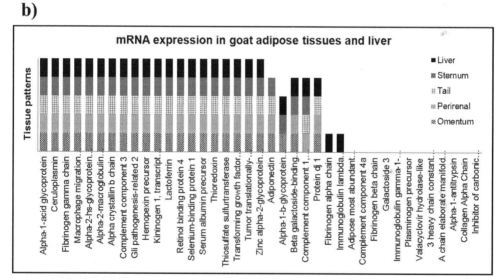

Figure 1. Results: the goat adipose tissue proteome and mRNA expression analysis. Following the pathways they are involved in, proteins were grouped in four major categories, based on similarity of functions: a1. The goat adipose tissue proteome, most of the 713 proteins are involved in metabolic processes a2. Classification of the proteins expressed only in subcutaneous adipose tissue, a3. Classification of the proteins expressed only in visceral adipose tissue. Proteomic data for some selected immune related proteins were validated by qualitative RT-PCR: b. mRNA expression of the selected proteins in adipose tissue and liver samples.

each adipocyte, the effective capability of AT to produce these immune related proteins was assessed by detection of the mRNA by qualitative PCR. 39 proteins of interest were selected (Table 1 and Table 2). PCRs generally confirmed the presence of the mRNAs in AT samples (Figure 1b). Only few genes, such as fibrinogen alpha chain and immunoglobulin gamma-1 and lambda light chain, weren't found in ATs samples, suggesting they were delivered from the blood. In some cases, due to the absence of the goat sequence, the bovine sequence was

Table 1. Proteins selected for further analysis, involved in immune and inflammatory response, expressed only in some of the adipose tissue deposits investigated.

Liver, omentum, sternum	Collagen alpha-1-chain
Omentum, sternum, tail	Fibrinogen beta chain
Omentum, perirenal, tail	Tumor translationally-controlled 1
Omentum, perirenal	Adiponectin; Fibrinogen alpha chain
Sternum, tail	3-heavy chain constant region; Galactoside 3
Omentum	Gli pathogenesis-related 2; Plasminogen precursor
Perirenal	Complement component 1, q subcomponent binding protein; Complement component 4a; Kininogen 1, isoform 1
Sternum	Transforming growth factor beta 68kda
Tail	Valacyclovir hydrolase-like; Zinc alpha-2-glycoprotein precursor

Table 2. Proteins selected for further analysis, involved in immune and inflammatory response, expressed in all adipose tissue samples.

Liver, omentum, perirenal, sternum, tail	
A chain elaborate manifold of short hydrogen bond arrays mediating binding of active site-directed serine protease inhibitors	Hemopexin precursor
	Immunoglobulin gamma-1-chain
	Immunoglobulin lambda light chain
Adipose most abundant gene transcript 2	Inhibitor of carbonic anhydrase-like
Alpha-1-acid glycoprotein	Lactoferrin L
Alpha-1-b-glycoprotein precursor	Macrophage migration inhibitory factor
Alpha-2-hs-glycoprotein precursor	Protein dj 1
Alpha-2-macroglobulin precursor	Retinol binding protein 4
Alpha crystallin b chain	Selenium-binding protein 1
Beta galactoside binding lectin precursor	Serum albumin precursor
Ceruloplasmin	Thioredoxin
Complement component 3	Thiosulfate sulfurtransferase
Fibrinogen gamma chain	

used for primer design. The absence of the mRNA of some genes in liver samples suggests that the bovine sequences used are not conserved between the two species. Results strongly suggest that most of the selected adipokines are expressed also in AT, confirming, in goats, the important role of AT in the immune and inflammatory response.

References

Conesa, A., Götz, S., García-Gómez, J.M., Terol, J., Talón, M. and Robles, M., 2005. Blast2GO: a universal tool for annotation, visualization and analysis in functional genomics research. Bioinformatics 21(18):3674-6.

Danielsen, M., Pedersen, L.J. and Bendixen, E., 2011. An *in vivo* characterization of colostrum protein uptake in porcine gut during early lactation. Journal of Proteomics 74(1):101-9.

Lemor, A., Mielenz, M., Altmann, M., von Borell, E. and Sauerwein, H., 2010. mRNA abundance of adiponectin and its receptors, leptin and visfatin and of G-protein coupled receptor 41 in five different fat depots from sheep. Journal of Animal Physiology and Animal Nutrition 94(5):e96-101.

Rajesh, R.V., Heo, G.N., Park, M.R., Nam, J.S., Kim, N.K., Yoon, D., Kim, T.H. and Lee, H.J., 2010. Proteomic analysis of bovine omental, subcutaneous and intramuscular preadipocytes during *in vitro* adipogenic differentiation. Comparative Biochemistry and Physiology – Part D: Genomics and Proteomics 5(3):234-44.

Thomas, P.D., Campbell, M.J., Kejariwal, A., Mi, H., Karlak, B., Daverman, R., Diemer, K., Muruganujan, A. and Narechania, A., 2003. PANTHER: a library of protein families and subfamilies indexed by function. Genome Research. 13(9):2129-41.

Comparison of the patterns of milk serum proteins in subclinical mastitis caused by four common pathogens in dairy cows

Shahabeddin Safi[1], Vahid Rabbani[2] and Seyed Hamed Shirazi-Beheshtiha[2]
[1]Department of Clinical Pathology, Faculty of Specialized Veterinary Sciences, Science and Research Branch, Islamic Azad University, Tehran, Iran; safishahab@yahoo.com
[2]Department of Clinical Sciences, Faculty of Veterinary Medicine, Karaj Branch, Islamic Azad University, Karaj, Iran

Introduction

At the present time, one of the diagnostic methods supplemented with somatic cell count and bacterial culture is measuring the acute phase proteins in milk samples from cows with subclinical mastitis. Acute phase proteins are group of proteins that their increase (positive acute phase proteins) or decrease (negative acute phase proteins) can be a sign of inflammation, trauma, or infections in the mammary glands (Murata *et al., 2004*). Those acute phase proteins in serum and milk which are separable by cellulose acetate electrophoresis include albumin, immunoglobulin, α-lactalbomin and β-lactoglobulin. Although there is a good body of knowledge on serum milk proteins but the comparison of the effects of different pathogens causing subclinical mastitis on the patterns and changes of APPs in milk serum has not been studied yet. The objective of this study was to compare the increase or decrease patterns of serum milk APPs during subclinical mastitis caused by *Streptococcus agalactiae*, *Staphylococcus aureus*, *Streptococcus uberis* and *Corynebacterium bovis* and also to determine the correlation coefficients of these proteins with somatic cell counts as the gold standard of subclinical mastitis.

Material and methods

From February to October 2010, 45 Holstein cows were selected randomly from 5 dairy farms in Tehran province, Iran. Clinical examination was performed on each cow before obtaining the samples. All cows chosen for this study were clinically normal at the time of sampling. Cows with clinical signs such as dullness or anorexia and with the clots in the milk (mild mastitis) were excluded from the study. Cows in late pregnancy or early lactation also were excluded from the study.

Quarters with SCC greater than 100,000 cells/ml and positive bacterial culture result were considered as affected. The opposite diagonal quarter were also sampled and were considered as intraanimal control if its SCC was below 100,000 cells/ml and had negative bacterial culture result (Table 1). The pathogens which were studied included *S. agalactiae*, *S. aureus*, *C. bovis*, and *S. uberis* which are amongst common pathogens causing subclinical mastitis in Iran

Table 1. SCC and different APP percentages in milk serums of different pathogens causing subclinical mastitis compared to diagonally opposite healthy quarter. The cut-off point for discrimination of healthy quarters and affected ones was considered as 100,000 cells/ml.

Quarters	SCC (100) cells/ml	Albumin %	α-lactalbumin%	β-lactoglobulin%	Immunoglobulin%
Healthy	<100	4.2	34.02	54.47	8.32
Streptococcus agalactiae	>100	11.7**	27.46	31.32*	30**
Healthy	<100	3.48	36.6	50.72	9.22
Streptococcus uberis	>100	9**	30.65	32.7**	26.6*
Healthy	<100	4.43	32.68	51	11.8
Staphylococcus aureus	>100	7.24*	26.29*	37.15**	29.32**
Healthy	<100	3.46	31.21	54.2	11
Corynebacterium bovis	>100	8.5**	34.38	37.9**	19.25*

* = P-value <0.05; ** = P-value <0.01.

(*Bolourchi et al., 2008*). 11 quarters with *S. agalactiae*, 11 quarters with *S. uberis*, 12 quarters with *S. aureus*, and 11 quarters with *C. bovis* were cultured. The milk sample was used for an automated SCC using a Fossomatic 90 analyzer (Foss Electric, Hillerød, Denmark) and blood agar and McConkey agar media were used for routine bacteriologic examination according to National Mastitis Protocol.

Milk serums were separated according to *Ishikawa et al.*, (1982). In order to separate serum milk proteins acetate cellulose electrophoresis was chosen as a simple routine method which can be used in most laboratories. The samples were placed on an acetate cellulose and electrophoresed for 15 minutes at 180 volts in a barbital buffer at pH 8.6 using Helena laboratories, Beaumont, Texas, USA. The separated bands were quantified using densitometry at 525 nm and the concentration of each band was measured. The concentration of total serum protein in milk was determined by biuret method and the concentration of protein in each band was calculated.

The bacteriologic results from the cows in this study with subclinical mastitis reflected the usual pathogenic bacteria, *S. agalactiae* and *S. aureus*, isolated from quarters affected with subclinical mastitis in Iran.

Results and discussion

The bacteriologic results from the cows in this study with subclinical mastitis reflected the usual pathogenic bacteria isolated from quarters affected with subclinical mastitis in Iran.

Figure 1 Panel A, shows the separated bands from serum milk of cows with subclinical mastitis caused by *S. agalactiae, S. aureus, S. uberis,* and *C. bovis* and electrophoretic pattern of normal serum milk.

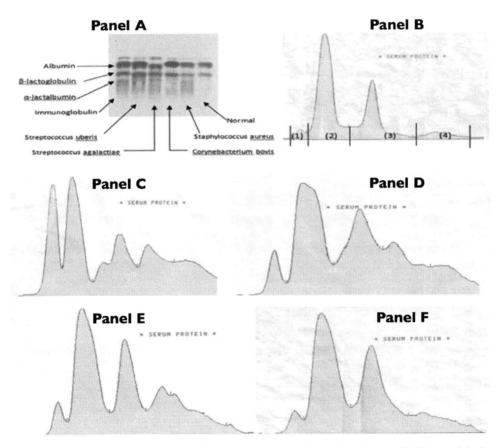

Figure 1. Electrophoretic pattern of milk serum in a normal cow. (1) albumin; (2) β-lactoglobulin; (3) α-lactalbomin; and (4) immunoglobulin and cows with S. agalactiae, S. uberis, S. aureus, C. bovis *infection.*

Figures 1 panel B to E show the electrophoretic patterns of subclinical mastitis caused by *S. agalatiae, S. uberis, S. aureus,* and *C. bovis,* respectively.

Conclusion

The results of the present study showed that the studied pathogens cause different electrophoretic patterns in milk of the affected quarters. The regression of the studied APPs and the related SCCs can be influenced by these pathogens as well. So it can be concluded that the electrophoresis of serum milk samples could be used to differentiate the pathogens causing subclinical mastitis.

References

Bolourchi, M., Mokhber Dezfouli, M.R., Kasravi R., Moghimi Esfandabadi, A. and Hovareshti, P., 2008. An estimation of national average of milk somatic cell count and production losses due to subclinical mastitis in commercial dairy herds. Iranian Journal of Veterinary Research 63:263-266.

Ishikawa, H., Shimizu, T., Hirano, H., Saito, N. and Nakano, T., 1982. Protein composition of whey from subclinical mastitis and effect of treatment with levamisole. Journal of Dairy Science 65: 653-658.

Murata, H., Shimada, N., Yoshioka, M., 2004. Current research on acute phase proteins in veterinary diagnosis: an overview. Veterinary Journal 168: 28-40.

National Mastitis Council, 1990. Microbiological procedures for the diagnosis of bovine udder infection. 3rd Ed. National Mastitis Council, Inc. Arlington, VA, USA.

Intrauterine treated lambs as a model for the study of Intrauterine Growth Restriction (IUGR) and consequent metabolic disorders

Tommaso Serchi[1], Jane Robinson[2], Neil Evans[2], Jenny Renaut[1], Lucien Hoffmann[1] and Arno C. Gutleb[1]

[1]*Département Environnement et Agro-biotechnologies, Centre de Recherche Public – Gabriel Lippmann, Belvaux, GD Luxembourg; serchi@lippmann.lu*
[2]*Institute of Biodiversity, Animal Health and Comparative Medicine, College of Medical, Veterinary and Life Sciences, University of Glasgow, Glasgow, G61 1QH, Scotland, United Kingdom*

Objectives

Developmental exposure to environmental and therapeutic chemicals, including excess androgens, has been reported to induce Intrauterine Growth Restriction (IUGR) and abnormal post-natal growth. It has now been clearly shown that IUGR and subsequent catch-up growth can result in a life-long vulnerability to diseases/syndromes that affect both the quality and duration of life. Ovine foetuses that are exposed to exogenous steroids exhibit IUGR, abnormal post-natal growth trajectories and altered metabolic and reproductive function and thus provide a model to examine the programming of conditions such as metabolic syndrome that has recently attracted much scientific and societal interest (Eckel *et al.*, 2005). Metabolic syndrome encompasses a group of risk factors that occur together and increase the risk of coronary artery disease, stroke and diabetes type 2, all of which increase the financial burden on health systems world-wide.

The aim of this project is to determine the relative contributions of, and mechanisms through which changes in the pancreatic proteome contribute to IUGR, the abnormal growth trajectory, and later life predisposition to conditions such as 'metabolic syndrome' seen in sheep exposed to androgens during foetal life.

Animals, material and methods

Animals and ethical approval

Poll Dorset ewes were born and maintained at the University of Glasgow farm (55°55'N). Studies on the reproductive, growth, and metabolic axes were approved by the university's Welfare and Ethics Committee and carried out under Home Office License PPL 60/3485.

Reproductive cycles of the ewes were synchronized using an intravaginal CIDR device (Inter Ag, New Zealand), and the ewes were mated to a Poll Dorset ram. Mating was monitored by fitting the ram with a marking harness. Pregnancy was assessed by ultrasound at about 60 days post-mating.

Before confirmation of pregnancy, ewes were randomly allocated to treatment groups to receive testosterone propionate (TP), dihydro-testosterone (DHT), or no injections as follows: of 48 pregnant ewes, 32 were given twice-weekly intramuscular injections of 100 mg of TP (n=17) or DHT (n=15) in vegetable oil. Injections began at day 30 after mating and continued until day 90 of pregnancy (term, 147 days; (Robinson *et al.*, 2012)). Sixteen pregnant ewes served as controls and did not receive any injections. Data from a similar study in sheep have shown that this treatment regime raises androgen concentrations in female fetuses to those similar to males (Veiga-Lopez *et al.*, 2011). Lambs were euthanized (Somulose, 1 ml/kg body weight, Dechra Veterinary Products, Shrewsbury, United Kingdom) at 10-11months of age, following puberty (at approximately 7 months of age). The pancreas was immediately removed, snap frozen and stored at -80 °C until further analyses. Other tissues were collected at the same time. Thirty lambs were randomly chosen (5 from each treatment, controls, TP and DHT, and gender) for the proteomic analysis.

2D-Difference in Gel Electrophoresis (2D-DiGE), protein identification and analysis

Proteins from lambs' pancreas were extracted following an optimized protocol that had been established earlier, with minor modifications (Kaulmann *et al.*, 2012). Extraction of proteins was achieved by addition of lysis buffer 7 M Urea, 2 M Thiourea, 2% CHAPS, 40 mM Tris (pH 8.5) (Sigma Aldrich, Lyon, France) and protease inhibitor (GE Healthcare, Uppsala, Sweden) to the frozen tissue. Solubilization of proteins was favored by mechanical grinding followed by 5 cycles of 30 seconds sonication and 30 seconds ice. The solution containing the proteins was then clarified by 10 minutes centrifugation at 14,000 rpm. Protein concentration was determined using the Bradford method reading the absorbance at 595 nm and comparing it with the internal standard curve obtained with serial dilutions of a 2 mg/ml bovine serum albumin standard.

For the labeling, 50 μg of proteins for each sample were labeled with either 400 pmol Cy3 or Cy5 dyes. In addition, a pool containing equal amounts of each sample was labeled with Cy2 dye in order to be used as internal standard. The first-dimension isoelectric focusing was performed using an IpgPhore III (BioRad, Hercules, CA, USA) on 24cm strips (GE Healthcare, Uppsala, Sweden), with a 3-10 linear pH gradient. The second-dimension separation was performed on a 12.5% pre-cast polyacrylamide gel (Serva, Heidelberg, Germany), on a High Performance Electrophoresis system (Serva, Heidelberg, Germany). Images were acquired with a scanner Typhoon 9400 (GE Healthcare, Uppsala, Sweden) at a spatial resolution of 100 μm. Obtained images were then analyzed using Decyder 2D software 7.0 (GE healthcare, Uppsala, Sweden).

The study is still ongoing and proteins identification will be achieved by MALDI-TOF/TOF (5800 MALDI-Tof/Tof AbiSciex) mass spectrometry.

Results and discussion

Both male and female animals exposed to TP or to DHT, between days 30 and 90 of a 147 day gestation, are significantly lighter at birth compared to control animals (Figure 1A). Both steroid treated groups exhibit catch-up growth during the post-natal period, however, animals exposed to TP also exhibit overshoot growth and were significantly heavier than control and DHT exposed animals by 11 months of age (Figure 1B). The differential growth patterns documented suggest that catch-up growth is programmed by both oestrogen and androgen receptor activity, while overshoot growth is programmed by an *in utero* action of testosterone solely via the oestrogen receptor. Relative to programmed metabolic dysfunction, foetuses exposed to androgens during this critical developmental window also develop hyperinsulinaemia and insulin resistance, which are only evident after puberty (Figure 1C). Moreover in the TP and DHT treated females evident signs of masculinization were present (Lamm *et al.*, 2012).

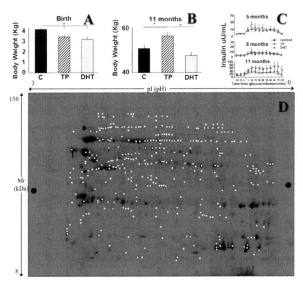

Figure 1. Effects of intrauterine TP and DHT treatment on lambs. Panel A).Weights of controls (C), testosterone propionate (TP) treated and dihydro-testosterone (DHT) treated Lambs at birth. Panel B) Weights of controls (C), testosterone propionate (TP) treated and dihydro-testosterone (DHT) treated lambs at 11months of age. Panel C) Plasma insulin levels following a bolus i.v. glucose injection (1mmol/kh 1wt) in control testosterone propionate (TP) treated and dihydro-testosterone (DHT) treated ewes during pre-pubertal (5 months), peripubertal (8 months) and post pubertal (11 months) periods. Panel D) A representative 2D gel. Proteins were separated in 1st dimension using 24 cm strips holding a 3-10 non-linear pH gradient and in 2nd dimension using 12.5% polyacrylamide large format pre-cast gels. Proteins of interested which were picked and will be subjected to identification are shown in orange.

From the preliminary analysis of the proteomic data many differences between genders and in reaction to the treatment can be highlighted. Proteins profile for each animal and group was analysed by principal component analysis and hierarchical clustering. In general, the male and female controls are very well separated by their protein pattern. The treated male lambs cluster close to their relative control and the proteins profile does not differ too much from the untreated lambs. In the case of the TP and DHT treated females, these groups seem to be much more similar to the male group and the masculinization effect appears clear. Detailed data on protein expression are not provided here since the identifications are still on-going and will be available in short time. A representative gel, with highlighted the picking locations is showed in Figure 1D.

It is clear that exposure to androgens during critical periods of *in utero* development alters the physiological regulatory systems responsible for growth e.g. tissue development, muscle growth, adiposity and energy regulation/partitioning, even though the exact mechanism is still to be clarified.

References

Eckel, R.H., Grundy, S.M. and Zimmet, P.Z., 2005. The metabolic syndrome. Lancet 365: 1415-1428.

Kaulmann, A., Serchi, T., Renaut, J., Hoffmann, L. and Bohn, T., 2012. Carotenoid exposure of Caco-2 intestinal epithelial cells did not affect selected inflammatory markers but altered their proteomic response. British Journal of Nutrition 108: 963-973.

Lamm, C.G., Hastie, P.M., Evans, N.P. and Robinson, J.E., 2012. Masculinization of the Distal Tubular and External Genitalia in Female Sheep With Prenatal Androgen Exposure. Veterinary Pathology 49: 546-551.

Robinson, J.E., Hastie, P.M., Shah, A., Smith, A. and Evans, N.P., 2012. Developmental Programming: Prenatal Androgen Exposure Alters the Gonadotroph Population of the Ovine Pituitary Gland. Journal of Neuroendocrinology 24: 434-442.

Veiga-Lopez, A., Steckler, T.L., Abbott, D.H., Welch, K.B., MohanKumar, P.S., Phillips, D.J., Refsal, K. and Padmanabhan, V., 2011. Developmental Programming: Impact of Excess Prenatal Testosterone on Intrauterine Fetal Endocrine Milieu and Growth in Sheep. Biology of Reproduction 84: 87-96.

Effect of estradiol on biochemical bone metabolism markers in dairy cows

Jože Starič[1], Jožica Ježek[1], Ivica Avberšek Lužnik[2], Martina Klinkon[1], Blaž Krhin[3] and Tomaž Zadnik[1]
[1]*Clinic for ruminants, Veterinary faculty, University of Ljubljana, Ljubljana, Slovenia;*
joze.staric@vf.uni-lj.si
[2]*Splošna bolnišnica Jesenice, Jesenice, Slovenia*
[3]*University medical center Ljublajna, Ljubljana, Slovenia*

Objectives

Biochemical markers of bone metabolism (BBM) are products of bone tissue formation and resorption that escape to blood and urine where we can measure them. Biochemical markers of bone tissue formation (for instance bone alkaline phosphatase (bALP)) are products of osteoblast activity and biochemical markers of bone tissue resorption (for instance C terminal telopeptide crosslinks of collagen 1 (CTx)) are by products of osteoclast activity. Markers of bone metabolism can indicate if bone metabolism is more anabolic, catabolic or in general more or less active (Allen, 2003; Christenson, 1997). In high producing dairy cows there are tremendous changes in Ca requirements in the so called transition period when cows go from dry period into lactation. We expect increased catabolic bone metabolism at the beginning of lactation as absorbable Ca demand suddenly increases from about 20 g to around 60 g or more per day in high yielding cows because of colostrum and milk production concurrent with a relatively low dry matter (nutrient) intake. A significant proportion of dairy cows are not successful in maintaining normocalcaemia at the beginning of lactation and they develop subclinical hypocalcaemia with all negative effects on health, welfare and production or progress toclinical hypocalcaemia, so called milk fever (Goff, 2000). BBM have been detected in dairy cattle (Filipovic *et al.*, 2008; Holtenius and Ekelund, 2005; Iwama, 2004; Liesegang *et al.*, 2000; Starič, 2012). Studies suggest that BBM can be used as a prepartum tool to predict if a cow is going to have milk fever after calving (Starič, 2010).

It is well known that estradiol has osteoprotective influence on bone tissue (Christenson, 1997). Its blood concentration is known to reach peack values just before calving (Patel *et al.*, 1999). We expect that it also influences values of BBM.

The aim of the study was to establish if BBM are associated with blood estradiol concentration in dairy cows from dry period until 20 days in lactation.

Material and methods

The study was conducted at a dairy farm with intensive milk production, with a loose housing system based on cubicles and pasture during warm seasons of the year. Forty-one clinically healthy cows before at least the 4th lactation were included in the study, 20 during winter time and 21 during summer time. During wintertime the animals were indoors all the time, while during summer time they were also on pasture every day. Cows were fed the usual winter ration based on home produced forages: grass silage, maize silage, hay and straw according to NRC (2001) recommendations. No anion salts were added to the diet and DCAD of the ration ranged from +200 to +300 mEq/kg dry matter. We monitored and obtained blood samples from investigated animals in different physiological periods during periparturient period in four sampling sessions:

1. 1 month before calving,
2. 10 or less days before calving,
3. within 48 hours after calving and
4. 10 to 20 days in milk.

Venous blood samples were collected in evacuated tubes (10 ml) without any additives (Venoject, plain silicone coated, Terumo Europe N.V., Belgium) from v. caudalis mediana according to the protocol between 9 and 11 a.m., to avoid daily fluctuations in analytes. After blood clotting, samples were centrifuged at 3000 rpm for 10 minutes then supernatants were centrifuged again at 3,000 rpm for 10 minutes at room temperature. Harvested blood serum was stored at -20 °C until analysis.

Blood serum bALP activity was measured using Alkphase-B kit (Metra Biosystems, USA) by enzyme immunoanalysis according to manufacturers' instructions on an Immulite 2500 analyser (Siemens, Germany). The absorbance at the end of reaction was measured with optical reader Humareader (Human, Egypt) at 405 nm wave length. CTx concentration in blood serum was measured by electrochemiluminiscent immunoanalysis ECLIA. The test was conducted using Elecysis 3 – CrossLaps kit on Elecys 1010 analyser (Roche Diagnostics, USA) according to manufacturers' instructions. 17β-estradiol was measured using 'Immulite 2000 – Estradiol' kit (Siemens Medical Solutions Diagnostics; Los Angeles, USA) on 'Immulite 2000' analyser (Siemens Medical Solutions Diagnostics; Los Angeles, USA) according to manufacturers' instructions.

Obtained values of investigated blood biochemical parameters were statistically analysed using descriptive statistics and correlations (Pearsons' or Kendalls' in case of nonparametric analysis). All the values that did not distribute normally were normalized. Analysis of variance with repetition was employed for testing influence of season and physiological period. If the factor value was not distributing normally nonparametric Fridman's test was used. Multiple comparisons as *post hoc* analysis were performed using Bonferroni's correction of *P*-value or in case of not normal distribution of data with multiple Wilcox's tests of ranks with Bonferronis' correction of α error. Statistical significance was set at $P<0.05$.

Results and discussion

Season did not influence values of BBM and estradiol while physiological period did (Table 1). In summer and winter group bALP is statistically significantly lower in samplings 1 and 2 (before calving) compared samplings 3 and 4 (after calving) and in winter group it is also higher in sampling 3 compared to sampling 4. CTx is higher in samplings 3 and 4 (after calving) compared to samplings 1 and 2 (before calving). Statistically significantly the highest CTx was at sampling 4 (10 to 20 days in milk). Values of estradiol in summer and winter group were higher at sampling 1 and 2 than in sampling 3 and 4. Statistically significantly the lowest value of estradiol in summer and winter group was in sampling 4 (10 to 20 days in milk).

The influence of estrogen on BBM was tested on combined summer and winter groups of cows to obtain a more representative sample and due to insignificant difference of values between summer and winter groups. As our data did not distribute normally, we used nonparametric method and interpret Kendalls' correlation coefficient. At sampling 1, 3 and 4 we could not confirm statistically significant correlations between estradiol and BBM. At sampling 2 estradiol statistically significantly positively correlated with bALP ($τ=0.23$; $P<0.05$).

Positive correlation between estradiol and bALP confirms estradiols' positive effect on bone tissue preservation in cattle. Results of the study suggest that high levels of estradiol in close prepartum time can add to development of hypocalcemia and milk fever because of an anabolic response of bone metabolism during the period of Ca shortage.

Table 1. Mean and standard deviation of bALP, CTx and estradiol in different sampling sessions during winter and summer time.

Sampling	bALP		CTx		Estradiol	
	Summer	Winter	Summer	Winter	Summer	Winter
1	12.10±3.38	15.38±4.36	0.097±0.084	0.223±0.165	0.262±0.202	0.269±0.206
2	15.09±7.69	16.96±9.01	0.135±0.145	0.143±0.077	1.111±0.865	2.058±2.753
3	25.71±10.15	31.02±7.22	0.625±0.400	0.623±0.325	0.480±0.750	0.361±0.717
4	29.23±7.26	24.26±5.11	1.1216±0.534	1.185±0.786	0.096±0.032	0.123±0.036

Acknowledgements

This work was financially supported by the Slovenian Research Agency; program P4-0092 (Animal health, environment and food safety).

References

Allen, M.J., 2003. Biochemical markers of bone metabolism in animals: uses and limitations. Vet Clin Pathol 32: 101-113.

Christenson, R.H., 1997. Biochemical markers of bone metabolism: an overview. Clin Biochem 30: 573-593.

Filipovic, N., Stojevic, Z., Zdelar-Tuk, M. and Kusec, V., 2008. Plasma parathyroid hormone-related peptide and bone metabolism in periparturient dairy cows. Acta Vet Hung 56: 235-244.

Goff, J.P., 2000. Pathophysiology of calcium and phosphorus disorders. Vet Clin North Am Food Anim Pract 16: 319-337, vii.

Holtenius, K. and Ekelund, A., 2005. Biochemical markers of bone turnover in the dairy cow during lactation and the dry period. Res Vet Sci 78: 17-19.

Iwama, Y., Kamiya, M., Tanaka, M. and Shioya, S., 2004. The change in dry matter intake, milk yield and bone turnover of primiparous cows as compared with multiparous cows during early lactation. Anim Sci J 75: 213-218.

Liesegang, A., Eicher, R., Sassi, M.L., Risteli, J., Kraenzlin, M., Riond, J.L. and Wanner, M., 2000. Biochemical markers of bone formation and resorption around parturition and during lactation in dairy cows with high and low standard milk yields. J Dairy Sci 83: 1773-1781.

Patel, O.V., Takenouchi, N., Takahashi, T., Hirako, M., Sasaki, N. and Domeki, I., 1999. Plasma oestrone and oestradiol concentrations throughout gestation in cattle: relationship to stage of gestation and fetal number. Res Vet Sci 66: 129-133.

Starič, J. and Zadnik, T., 2010. Biochemical markers of bone metabolism in dairy cows with milk fever. Acta Vet (Beogr.) 60: 401-410.

Starič, J., Nemec, M., Zadnik, T., 2012. Bone metabolism markers and blood minerals concentration in dairy cattle during pregnancy and lactation. Slov Vet Res 49: 193-200.

Dynamics of metabolic and oxidative stress parameters in dairy heifers during transition period

Petra Zrimšek[1], Ožbalt Podpečan[1,2], Janko Mrkun[1], Marjan Kosec[1], Zlata Flegar-Meštrić[3], Sonja Perkov[3], Jože Starič[4], Mirna Robić[5], Maja Belić[5] and Romana Turk[5]

[1]Clinic for Reproduction and Horses, Veterinary Faculty, University of Ljubljana, Ljubljana, Slovenia; petra.zrimsek@vf.uni-lj.si
[2]Savinian Veterinary Policlinic, Žalec, Slovenia
[3]Clinical Institute of Medical Biochemistry and Laboratory Medicine, Clinical Hospital 'Merkur', Zagreb, Croatia
[4]Clinic for Ruminants, Veterinary Faculty, University of Ljubljana, Ljubljana, Slovenia
[5]Department of Pathophysiology, Faculty of Veterinary Medicine, University of Zagreb, 10000 Zagreb, Croatia

Objectives

Dairy cows in early postpartum period experience a rapid increase in milk yield, whereas animal feed consumption ability rises slowly and can not follow increased needs for nutrients (Eicher, 2004). Therefore dairy cows enter a period of negative energy balance (NEB) and metabolic stress. NEB is associated with the change in body condition score (BCS) and the rise of some metabolites such as non-esterified fatty acids (NEFA) and beta- hydroxybutyrate (BHB) (Lassen and Fettman, 2004). Intensified energy metabolism during periparturient period is accompanied by enlarged rates of reactive oxygen species (ROS) production and may results in oxidative stress. Calving causes significant, but temporary, changes in the antioxidant system of cows' blood (Gaal et al., 2006). Total antioxidant status (TAS) is a single measure that may effectively describe the dynamic equilibrium between pro-oxidants and antioxidant in plasma compartment (Ghiselli et al., 2000). Paroxonase-1 (PON1) is recognized as an important part of the mammalian natural anti-oxidative system (Mackness et al., 1991; Turk et al., 2008).

Possible relationship between lipid mobilisation indicators and oxidative stress markers during transition period of dairy heifers was investigated in the present study. The aim of our work was to evaluate the dynamic of metabolic and oxidative stress parameters (NEFA, BHB, TAS and PON1) in dairy heifers during transition period. Comparison of mentioned parameters from one week before to 8 weeks after calving with the values at insemination will contribute to the establishment of physiological ranges for biomarkers of oxidative stress in ruminant medicine.

Material and methods

Nineteen Holstein-Frisian dairy heifers, BVD virus and IBR/IPV virus free were included in the study from June 2011 to August 2012. Blood samples were collected at the time of insemination, one week before calving, at calving and 1, 2, 4 and 8 weeks post partum. The sera and plasma were separated, aliquoted and stored at -80 °C.

Serum BHB and NEFA were measured by the automated clinical chemistry analyzer RX Daytona (Randox Laboratories Ltd, Crumlin, UK) using commercially available kits RANBUT kit (Randox Laboratories Ltd, Crumlin, UK) and NEFA-HR kit (Walko Chemicals GmbH, Neuss, Germany) for BHB and NEFA determination, respectively following manufacturer's instructions.

PON1 activity was assayed by the slightly modified method of hydrolysis of paraoxon (Schiavon *et al.*, 1996) using Beckman Coulter AU 680 (Beckman Coulter Biomedical Ltd., Ireland). PON1 activity was expressed in international units (U/l) as the amount of substrate (paraoxon (O,O-diethyl-O-p-nitrophenylphosphate, Sigma Chemical Co, London, UK) hydrolyzed per minute and per litre of serum (μmol/min/l).

Plasma samples were assayed for total antioxidant capacity by an automated clinical chemistry analyser RX Daytona (Randox Laboratories Ltd, Crumlin, UK), using commercially available Total Antioxidant Status (TAS) kit (Randox Laboratories Ltd,, Crumlin, UK) using Trolox as standard (6-hydroxy-2,5,7,-tetramethylchroman-2-carboxylic acid). The results are expressed as mmol/l of Trolox equivalents.

Results and discussion

Dairy cows enter a period of negative energy balance (NEB) around calving, which leads to mobilization of body reserves, mainly fat, to balance the deficit between food energy intake and production requirements. Lipid mobilisation syndrome is characterized by an increase in NEFA and BHB (van Knegsel *et al.*, 2005).

Our results show that NEFA concentrations increased significantly at calving and remain increased until 4 weeks postpartum indicating adipose tissue breakdown (Butler, 2000). BHB showed similar pattern as NEFA (Figure 1A), although the highest NEFA and BHB concentrations were observed 2 and 4 weeks postpartum, respectively. BHB is used for milk synthesis as a part of ketone bodies which provide an important form of energy to peripheral tissues in case when carbohydrate levels are reduced (Duffield, 2000). Eight weeks postpartum NEFA concentrations decreased to the level observed at insemination.

TAS levels around insemination and from calving to 4 weeks post partum did not differ significantly (P>0.05), whereas one week before calving TAS was significantly lower than in all other samplings (*P*<0.05). The highest value of TAS was observed 8 weeks post partum (Figure 1B).

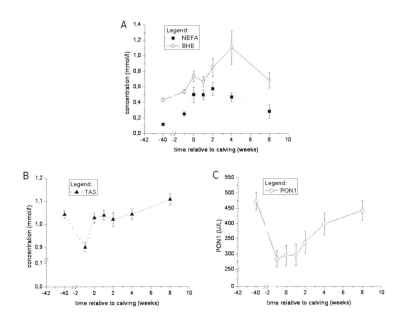

Figure 1. Patterns of NEFA, BHB (A) and TAS (B) concentrations and PON1 (C) activity (mean ± SE) in blood of heifers from 1 week before calving to 8 weeks post partum in comparison with level at insemination.

PON1 activity was significantly lower 1 week before parturition compared to the level at insemination ($P<0.05$) and remain at the lowest level up to one week after calving. Thereafter, the activity of PON1 continuously increased to 4 weeks post partum; 4 and 8 weeks post partum PON1 activity was not significantly different with the level at insemination ($P>0.05$) (Figure 1C). These results support the suggestion that calving causes significant, but temporary, changes in the antioxidant system of cows' blood (Gaal *et al.*, 2006). PON1 activity was also decreased during the transition period in our previous studies (Turk *et al.*, 2004, 2005, 2008). However, since PON1 was decreased before and after parturition, it could be a better parameter for oxidative stress and negative energy balance than TAS which concentration decreased only one week before parturition.

PON1 activity significantly negatively correlated with NEFA ($r=-0.490$, $P<0.001$) and BHB ($r=-0.392$, $P<0.001$) and positively with TAS ($r=0.443$, $P<0.001$). TAS correlated only with PON1, but not with NEFA or BHB what could support the prediction that PON1 activity is more influenced by energy deficit then TAS. Because of a considerable inter-individual variability of PON1 activity between animals, PON1 could be a valuable marker for a herd health management.

Transition period markedly affected concentrations of investigated parameters. Significant positive correlations between PON1 and TAS and negative ones between PON1 and NEFA as well as BHB suggest that in period of negative energy balance (NEB) in early postpartum period, lipid mobilisation and oxidative stress are dependent.

Acknowledgements

This work was supported by the Slovenian Ministry of Education, Science, Culture and Sport, programme groups "Endocrine, immune, nervous and enzyme responses in healthy and sick animals" (P4-0053) and "Animal health, environment and food safety" (P4-0092), Ministry of Science, Education and Sport Republic of Croatia (research project "Role of antioxidant enzymes and lipoproteins in reproduction of cows and sows (MZOŠ 053-0532265-2231) and bilateral project between Slovenia and Croatia (BI-HR/12-13-006).

References

Butler, W.R., 2000. Nutritional interactions with reproductive performance in dairy cattle. Anim Reprod Sci 60-61: 449-457.

Duffield, T., 2000. Subclinical ketosis in lactating dairy cattle. Vet Clin North Am Food Anim Pract 16: 231-253, v.

Eicher, R., 2004. Evaluation of the metabolic and nutritional situation in dairy herds diagnostic use of milk components. Med Vet Quebec 34: 36-38.

Gaal, T., Ribiczeyne-Szabo, P., Stadler, K., Jakus, J., Reiczigel, J., Kover, P., Mezes, M. and Sumeghy, L., 2006. Free radicals, lipid peroxidation and the antioxidant system in the blood of cows and newborn calves around calving. Comp Biochem Physiol B Biochem Mol Biol 143: 391-396.

Ghiselli, A., Serafini, M., Natella, F. and Scaccini, C., 2000. Total antioxidant capacity as a tool to assess redox status: critical view and experimental data. Free Radic Biol Med 29: 1106-1114.

Lassen, E.D. and Fettman, M.J., 2004. Laboratoty Evaluation of Lipids. Veterinary Hematology and Clinical Chemistry. Lippincott Williams & Wilkins, Baltimore, Maryland, USA: 421-431.

Mackness, M.I., Harty, D., Bhatnagar, D., Winocour, P.H., Arrol, S., Ishola, M. and Durrington, P.N., 1991. Serum paraoxonase activity in familial hypercholesterolaemia and insulin-dependent diabetes mellitus. Atherosclerosis 86: 193-199.

Schiavon, R., De Fanti, E., Giavarina, D., Biasioli, S., Cavalcanti, G. and Guidi, G., 1996. Serum paraoxonase activity is decreased in uremic patients. Clin Chim Acta 247: 71-80.

Turk, R., Juretic, D., Geres, D., Svetina, A., Turk, N. and Flegar-Mestric, Z., 2008. Influence of oxidative stress and metabolic adaptation on PON1 activity and MDA level in transition dairy cows. Anim Reprod Sci 108: 98-106.

Turk, R., Juretic, D., Geres, D., Turk, N., Rekic, B., Simeon-Rudolf, V. and Svetina, A., 2004. Serum paraoxonase activity and lipid parameters in the early postpartum period of dairy cows. Res Vet Sci 76: 57-61.

Turk, R., Juretic, D., Geres, D., Turk, N., Rekic, B., Simeon-Rudolf, V., Robic, M. and Svetina, A., 2005. Serum paraoxonase activity in dairy cows during pregnancy. Res Vet Sci 79: 15-18.

Van Knegsel, A.T., van den Brand, H., Dijkstra, J., Tamminga, S. and Kemp, B., 2005. Effect of dietary energy source on energy balance, production, metabolic disorders and reproduction in lactating dairy cattle. Reprod Nutr Dev 45: 665-688.

The proteomics of wool follicles

Jeffrey E. Plowman, Sivasangary Ganeshan, Joy L. Woods, Santanu Deb-Choudhury, David R. Scobie and Duane P. Harland
AgResearch Ltd, Private Bag 4749, Christchurch 8140, New Zealand;
jeff.plowman@agresearch.co.nz

Introduction

Messenger RNA hybridisation labelling of hair follicles has revealed that in both sheep and humans there is a sequential pattern to expression of the major keratin and keratin associated proteins (Langbein and Schweizer, 2005; Yu *et al.*, 2009). How expression sequence and post-translational changes relate to fibre development is largely unknown. Thus, this study was initiated to determine if proteomics could provide an insight into protein changes during the various stages of protein synthesis and keratinisation in the maturing wool fibre.

Follicle collection, preparation and analysis

A sample of skin was obtained from a recently slaughtered crossbred lamb at an abattoir, the wool removed with small animal clippers, and the skin cleaned with ethanol and distilled water. Wool follicles, dissected from the skin with a scalpel, were stored in phosphate buffered saline. Under a light microscope three distinct regions were observed in the lower follicle: the germinative matrix and elongation region of the follicle bulb and the opaque keratogenous region where most of the keratin and keratin associated proteins (KAP) are expressed (Marshall *et al.*, 1991). Samples were cut short of the translucent keratinisation zone where the final stage of intermediate filament assembly occurs during fibre hardening.

For the proteomic studies 10 follicles were selected and stored in a -80 °C freezer overnight. The proteins were extracted with lysis buffer and then digested with trypsin for 18 hours. Empore-assisted extraction of the peptides was then used prior to analysis on an AmaZon Speed ETD ion trap mass spectrometer (Bruker). The peak lists were searched against the NCBInr protein sequence database, augmented with an in-house expressed sequence tag database of sheep sequences using ProteinScape (Bruker). Proteins were also extracted from 30 follicles with the lysis buffer and separated by 2DE, in the first dimension on a 3-11 non-linear pH gradient and in a 7.5-17%T acrylamide gradient in the second dimension. The proteins in the gel were visualised with colloidal Coomassie Blue G250.

Protein composition of the follicle

A total of 93 proteins were identified in the digest from the wool follicle extract, of which two thirds were from the cell cytoplasm and nucleus. Of the trichocyte keratins, one cuticle Type II keratin (K82) was detected, along with six Type I and four Type II keratins from the fibre cortex (Table 1). Included among these were K35 and K85, which are also found in the cuticle. The KAPs were less well represented, with only six proteins out of a total number of 69 known sequences.

Also evident in the dissected follicle extract were a number of Type I and II epithelial keratins and trichohyalin, which most probably originate from the inner and outer root sheaths. Other keratin proteins were also observed, notably vimentin, a cytosolic protein, and lamin from the cell nucleus

One protein identified was desmoplakin, a component of functional desmosomes with the role of anchoring intermediate filament proteins to desmosomal plaques. Another desmosomal protein, plakoglobin was also observed.

The keratins in the 2DE map of the wool follicles showed a similar arrangement to those in the mature fibre with the Type I keratins appearing in a tight cluster and the Type II keratins forming a long train, both these exhibiting a shift to slightly higher pH regions. One significant difference was the presence of what appeared to be as many as 11 trains of spots in the Type I keratin region, as opposed to the four major strings in the mature fibre (Figure 1).

Future directions

Preliminary work has demonstrated that it is possible to further dissect the follicle into three parts based on the regions defined above. These are the proximal half of the follicle bulb up to about halfway up the dermal papilla (Auber's critical level), the remainder of the follicle bulb up to the keratogenous zone and the keratogenous zone itself. Work is currently focusing on extracting and identifying the proteins in these three zones.

Table 1. Keratin and related proteins identified in the wool follicles.

Protein type	Family	Proteins identified
Trichocyte keratins	Type I	K31, K33a, K33b, K34, K35, K38
	Type II	K81, K82, K83, K85, K86
	KAPs	KAP2.3, KAP3.2, KAP3.3, KAP7.1, KAP11.1, KAP13.1
Epithelial keratins	Type I	K14, K15, K18, K25, K27
	Type II	K1, K5, K6A, K7, K71,K73, K75
Proteins associated with keratins		Trichohyalin, desmoplakin
Other keratins		Vimentin, Lamin

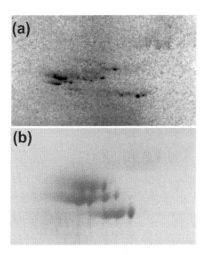

Figure 1. 2DE maps of the Type I keratin region in (a) wool follicles and (b) the mature fibre.

References

Langbein, L. and Schweizer, J., 2005. Keratins of the human hair follicle. Int Rev Cytol 243: 1-78.

Marshall, R.C., Orwin, D.F. and Gillespie, J.M., 1991. Structure and biochemistry of mammalian hard keratin. Electron Microsc Rev 4: 47-83.

Yu, Z., Gordon, S.W., Nixon, A.J., Bawden, C.S., Rogers, M.A., Wildermoth, J.E., Maqbool, N.J. and Pearson, A.J., 2009. Expression patterns of keratin intermediate filament and keratin associated protein genes in wool follicles. Differentiation 77: 307-316.

Relationships between mediators of the acute phase response and altered energetic metabolism in dairy cows after calving

Gabriel Kováč, Csilla Tóthová and Oskar Nagy
Clinic for Ruminants, University of Veterinary Medicine and Pharmacy, Košice, Slovakia;
kovac@uvm.sk

Objectives

Parturition, changes in homeostasis, metabolic and physiological challenges occuring in the period after calving may contribute to the activation of host immune system, including the initiation of the production of acute phase proteins. They have various activities by which contribute to germs destruction, they reduce tissue damage and help its regeneration (Eckersall and Bell, 2010). Major bovine acute phase proteins, haptoglobin and serum amyloid A, play an important role also in the reproductive processes (Krakowski and Zdzisińska, 2007). Moreover, the transition of pregnancy to lactation, with the concomitant negative energy balance, requires substantial adaptation of the cow, including metabolic and physiological adaptations, and is accompanied by changes in whole metabolism (Leroy *et al.*, 2008). However, the pathophysiology and the relationships between the aforementioned metabolic events and changes in immune functions in dairy cows are less well documented.

Therefore, this study was aimed at the evaluation of the relationships between the activated immune response, characterized by the presence of acute phase proteins, and altered energetic metabolism in dairy cows after calving.

Material and methods

Into the evaluation we included 195 clinically healthy dairy cows of a Slovak spotted breed from a conventional dairy farm. The monitored cows were in a period of 1-2 weeks after parturition, and showed no health disorders during the observation. The laboratory analyses were performed in blood samples collected by direct puncture of *v. jugularis*. Blood serum was analyzed for major bovine acute phase proteins – haptoglobin (Hp, mg/ml) and serum amyloid A (SAA, µg/ml), and selected variables of energetic profile – glucose (Glu, mmol/l), total cholesterol (TCH, mmol/l), total lipids (TL, g/l), triglycerides (TG, mmol/l), non-esterified fatty acids (NEFA, mmol/l), and β-hydroxybutyrate (BHB, mmol/l). Haptoglobin and SAA were assessed using commercial diagnostic kits (Tridelta Development, Ireland) in microplates. The reading of their absorbancies were performed on automatic microplate reader Opsys MR (Dynex Technologies, USA). The concentrations of Glu, TCH, TG, and BHB were determined using commercial diagnostic kits (Randox, United Kingdom) on automatic biochemical analyser ALIZE (Lisabio, France). Total lipids and NEFA were analyzed spectrophotometric method.

The obtained results from the evaluated cows were divided into two groups according to the measured concentrations of NEFA: Group A (n=108) – cows with serum concentrations of NEFA below 0.35 mmol/l; Group B (n=87) – cows with serum concentrations of NEFA above 0.35 mmol/l.

Statistical evaluation of the results was performed by assessment of average values (x) and standard deviations (SD). The significance of differences in measured values (P) between the groups of animals was evaluated by Mann-Whitney nonparametric test. Relationships between the concentrations of evaluated variables were calculated by linear regression and Spearman (R) correlations coefficient, including significance of the correlation.

Results and discussion

Analyses showed significantly higher mean serum concentrations of Hp and SAA in cows with concentrations of NEFA above 0.35 mmol/l (n=87) compared with those with serum NEFA concentrations below 0.35 mmol/l (n=108) ($P<0.001$ and $P<0.001$, respectively, Table 1). In serum concentrations of BHB we found a similar trend of significantly higher values in cows with NEFA concentrations above 0.35 mmol/l ($P<0.001$). On the other hand, cows with

Table 1. Comparison of the concentrations of Hp, SAA and selected parameters of energetic profile in two groups of dairy cows (x ± SD).

Variables	Group of cows		P-value
	A (n=108)	B (n=87)	
Hp (mg/ml)	0.067±0.113	0.607±0.610	<0.001
SAA (µg/ml)	30.77±20.95	93.94±43.64	<0.001
Glu (mmol/l)	4.19±0.57	3.95±0.44	<0.001
TCH (mmol/l)	3.25±1.11	3.05±1.12	n.s.
TL (g/l)	3.48±1.44	3.43±1.31	n.s.
TG (mmol/l)	0.14±0.11	0.12±0.09	n.s.
NEFA (mmol/l)	0.21±0.09	0.65±0.25	<0.001
BHB (mmol/l)	0.46±0.26	0.65±0.35	<0.001

Group A = cows with serum concentrations of NEFA below 0.35 mmol/l.

Group B = cows with serum concentrations of NEFA above 0.35 mmol/l.

P = significance of the differences of means between the groups of cows, n. s. = non significant.

higher values of NEFA showed significantly lower mean concentration of glucose ($P<0.001$). In mean concentrations of total cholesterol, total lipids, and triglycerides we observed no significant differences between the two groups of cows. The concentrations of both measured acute phase proteins – Hp and SAA significantly positively correlated with the values of NEFA (R=0.716, $P<0.001$; R=0.710, $P<0.001$, respectively), as well as BHB (R=0.291, $P<0.001$; R=0.300, $P<0.001$, respectively).

Parturition with following metabolic challenges constitutes a potentially stressful event for the dairy cow. One of the ways how an animal can manifest stress is in the form of activated acute phase response, including increased production of acute phase proteins by the liver. The physiological processes taking place around the time of parturition, especially increase in myometrial activity, involution of the uterus, as well as regeneration of the endometrium, may be also responsible for higher values of acute phase proteins (Rottmann, 2006). Moreover, homeostais of all the energy substrates and the whole metabolism is altered during the time around parturition. Hardardottir *et al.* (1994) reported also that the acute phase response initiated by processes occuring around parturition is associated with numerous changes in lipid and glucose metabolism, such as decreased cholesterol, accelerated lipolysis, and increased NEFA concentrations in plasma. Investigations in human medicine showed that altered lipid metabolism, increased concentrations of non-esterified fatty acids in blood serum, and oxidative stress may markedly influence the systemic inflammatory response, and the development of inflammatory-based diseases (Sordillo *et al.*, 2009).

On the other hand, the periparturient period is characterized by a sudden increase in energy requirements imposed by the onset of lactation and by a decrease in voluntary dry mater intake which results in negative energy balance (Leroy *et al.*, 2008). A significant adaptation to the aforementioned negative energy balance during the transition period is the mobilization of fat from body stores and the release of NEFA into the blood stream, which constitute important sources of energy in this period. Animals may react to these changes in metabolism with changes in the concentrations of some acute phase proteons. However, according to Sordillo *et al.* (2009) increased circulating NEFA concentrations are directly associated with increased systemic inflammatory conditions, and large amounts of adipose stores during time of energy deficiency are linked with adverse health effects on the transition cow. Other researches stated also that elevated NEFA concentrations are positive risk factors for many inflammatory periparturient diseases in dairy cows (Wood *et al.*, 2009). However, further studies are needed to explain the aforementioned relationships between the immune response, characterized by changes in the concentrations of acute phase proteins in dairy cows after calving, and between altered energetic metabolism.

Presented results indicate strong relationships between the variables of energetic metabolism, and between indicators of acute phase response in cows shortly after parturition. Understanding how all these metabolic factors interact with the immune system may help in developing disease control strategies that may aid in maintaining good health in dairy cattle.

Acknowledgements

This work was supported by the Slovak Research and Development Agency under contract No. APVV-0475-10 and by VEGA Scientific Grant No. 1/0592/12 from the Ministry of Education.

References

Eckersall, P.D. and Bell, R., 2010. Acute phase proteins: Biomarkers of infection and inflammation in veterinary medicine. Vet J 185: 23-27.

Hardardottir, I. Grunfeld, C. and Feingold, K.R., 1994. Effects of endotoxin and cytokines on lipid metabolism. Curr Opin Lipidol 5: 471-475.

Krakowski, L and Zdzisińska, B., 2007. Selected cytokines and acute phase proteins in heifers during the ovarian cycle course and in different pregnancy periods. Bull Vet Inst Pulawy 51: 31-36.

Leroy, J.I., Vanholder, T., van Knegsel, A.T., Garcia-Ispierto, I. and Bols, P.E, 2008. Nutrient prioritization in dairy cows early postpartum: mismatch between metabolism and fertility. Reprod Domestic Anim 43: 96-103.

Lippolis, J.D., 2009. Understanding the relationship between imunity and dairy cow production. Hoard's Dairy man 154: 481.

Rottmann, S., 2006. Einfluss von nichtsteroidalen Antiphlogistika (NSAID) auf hämatologische und klinisch-chemische parameter bei Rindern mit Dystokie. Inaugural-Dissertation, Veterinärmedizinische Fakultät, Leipzig, 113 pp.

Sordillo, L.M., Contreras, G.A. and Aitken, S.L., 2009. Metabolic factors affecting the inflammatory response of periparturient dairy cows. Anim Health Res Vet 10: 53-63.

Wood, I.G., Scott, H.A., Garg, M.I. and Gibson, P.G., 2009. Innate immune mechanisms linking non-esterified fatty acids and respiratory disease. Prog Lipid Res 48: 27-43.

Serum protein electrophoretic pattern in clinically healthy calves and cows

Csilla Tóthová, Oskar Nagy, Gabriel Kováč and Veronika Nagyová
Clinic for Ruminants, University of Veterinary Medicine and Pharmacy, Košice, Slovakia;
tothova@uvm.sk

Objectives

The determination of serum proteins and their electrophoretic profiles is an important diagnostic aid in clinical biochemistry. Serum protein electrophoresis is a common technique of laboratory diagnosis in human, and has been studied intensively in small animal and equine medicine (Riond *et al.*, 2009). However, in bovine medicine, the electrophoretic separation of serum proteins is a rarely used laboratory method. The pattern of serum protein electrophoresis results depends on the fractions of two major types of proteins: albumin and globulins. Many pathologic conditions can cause shifts in albumin and globulin concentrations (Keay and Doxey, 1982). Moreover, abnormalities of serum protein electrophoretic pattern must be interpreted in the view of the many other influences not associated with diseases. The age of evaluated animals is an important factor that may affect the concentrations of frequently analyzed biochemical variables, and may also influence the serum protein electrophoretic pattern.

The objective of this study was to determine the normal electrophoretic pattern of serum proteins in clinically healthy calves and cows using agarose gel electrophoresis, and to describe the possible age-dependent differences in serum protein fractions between young and adult cattle (Figure 1).

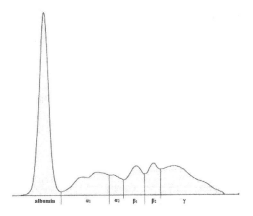

Figure 1. The electrophoretic pattern of serum proteins in clinically healthy calf.

Material and methods

Two different age groups of clinically healthy cattle (fourteen calves and thirteen adult dairy cows) from a conventional dairy farm were included into this study. The calves were of a Slovak spotted breed and its crossbreeds at the age of 4-6 months. The evaluated cows were of the same breed, and between 3-5 years of age. All the evaluated animals were in good general health without any obvious disease signs. Blood samples for the analyses were collected from both calves and cows by direct puncture of *v. jugularis* into serum gel separator tubes without anticoagulant. The harvested blood serum was analysed for total serum protein concentration and for the identification of serum protein fractions.

Total protein (TP, g/l) concentrations were assessed on an automated biochemical analyser Alizé (Lisabio, France) by the biuret method using commercial diagnostic kits (Randox). Serum protein fractions were separated by zone electrophoresis on a buffered agarose gel at pH 8.8 on an automated electrophoresis system Hydrasys (Sebia Corporate, France) using commercial diagnostic kits Hydragel 7 Proteine (Sebia Corporate, France) according to the procedure described by the manufacturer. The electrophoretic gels were scanned, and the serum protein fractions were visualized and displayed on densitometry system Epson Perfection V700 (Epson America Inc., California, USA) by light transmission and automatic convertion into an optical density curve presentation. Protein fractions were identified and quantified by computer software Phoresis version 5.50 (Sebia Corporate, France), and if necessary, corrected by visual inspection of the electrophoretogram. The relative concentrations (%) of the protein fractions were determined as the percentage of the optical absorbance, and the absolute concentrations (g/l) were calculated from the total serum protein concentrations. Albumin:globulin ratios (A/G) were computed from the electrophoretic scan.

Arithmetic means (x) and standard deviations (SD) calculated using descriptive statistical procedures. Mann-Whitney non-parametric test was used to compare the results and to evaluate the significance of differences in measured values between calves and cows.

Results and discussion

Serum protein electrophoresis on agarose gel identified in calves, as well as in cows, albumin, two α-globulin (α_1 and α_2), two β-globulin (β_1 and β_2), and γ-globulin fractions. There are very few reports published about more detailed serum protein profiles in cattle, and the data were obtained some years ago. Keay and Doxey (1982) presented only electrophoretograms with the number and magnitude of protein fractions, without determination of relative concentrations and calculating of absolute values for each fraction. Jawor *et al.* (2008) separated serum proteins in adult female cattle 4 fractions: albumin, α-, β-, and γ-globulins. These variations in the concentrations of serum protein fractions, reported by various authors might be caused by the used support media and electrophoretic techniques.

Albumin was the most prominent fraction in both young and adult animals, and constituted in average 48.3% and 42.3% of total serum proteins in calves and cows, respectively (Table 1). The results found in calves and cows differed significantly ($P<0.05$). Significant differences between the evaluated groups of animals we found also for α_1-globulins, with higher relative concentrations in calves compared with cows ($P<0.001$). Jawor *et al.* (2008) reported that serious inflammatory conditions are associated with higher concentrations of α-globulins, and that this increase is caused by the fact that the majority of acute phase proteins (haptoglobin, ceruloplasmin, α_1-acid glycoprotein, α_1-antitrypsin) occurs in this fraction. On the other hand, higher concentrations of α-globulins in calves are not necessarily a sign of the activation of inflammatory processes, or a sign of a disease. Higher values of α_1-globulins in calves may be related to the exposure of animals to changing nutritional and rearing factors, and may be associated with the normal process of growth.

Presented study indicated in calves significantly higher relative concentrations also for β_1-globulins ($P<0.05$)., which may be a result of suckling. An opposite trend we observed in the concentrations of γ-globulins. The relative concentrations of this fraction in calves were

Table 1. The concentrations of total serum proteins, serum protein fractions and A/G ratios in clinically healthy calves and dairy cows.[1]

Parameter		Group of animals	
		Calves (n=14)	Cows (n=13)
TP	g/l	71.7±3.65	83.2±8.87***
Alb	g/l	34.6±3.33	34.8±4.13
	%	48.3±3.47	42.3±6.83*
α_1-globulins	g/l	10.5±1.40	9.3±1.28
	%	14.6±2.06	11.2±1.27***
α_2-globulins	g/l	3.2±0.53	4.9±1.49**
	%	4.4±0.72	5.8±1.39*
β_1-globulins	g/l	6.0±0.60	6.2±1.28
	%	8.4±0.81	7.4±1.08*
β_2-globulins	g/l	5.8±1.47	6.8±1.38
	%	8.1±2.04	8.2±1.16
γ-globulins	g/l	11.7±2.14	21.1±6.31***
	%	16.2±2.74	25.1±5.65***
A/G	g/l	0.94±0.12	0.76±0.21*

[1] Asterisks in rows mean statistical significance of differences between calves and cows: * = $P<0.05$; ** = $P<0.01$; *** = $P<0.001$.

significantly lower than the values measured in cows (*P*<0.001). Chaudhary *et al.* (2003) suggested also that lower concentrations of γ-globulins in calves are caused by the immaturity of the lymphoid system, and that these lower values remain until the production of globulins by the maturing immune system. Therefore, lower relative concentrations of γ-globulins in calves compared with adult animals observed also in our study may be connected to the normal process of growth and developing immune system in calves. Presented results suggest a marked shift in the concentrations of albumin and globulins in calves compared with cows (higher relative concentrations of albumin and lower percentages of γ-globulins in calves), presumably caused by changing globulin patterns during development. Age is thus an important consideration in the interpretation of results of serum protein electrophoresis.

The albumin/globulin ratio is of special interest for clinicians because it allows systematic classification of the electrophoretic profile and identification of dysproteinemias. In our study, presented data showed significantly lower A/G ratio in cows compared with young animals, as a result of lower albumin and higher globulin concentrations in adult animals. Therefore, the A/G ratio must be interpreted cautiously with attention paid to which part of the ratio has changed.

In conclusion, presented study provides preliminary results for the determination of physiologic concentrations of bovine serum protein fractions separated by agarose gel electrophoresis. Seeing that the concentrations of serum protein fractions have not been previously sufficiently standardized in cattle, the obtained data would be useful for clinicians by the determination of dysproteinemias and the evaluation of various pathological conditions, providing a basis for further specific laboratory investigations. Moreover, in our study presented results showed a marked influence of age on the concentrations of several serum protein fractions.

Acknowledgements

This work was supported by the Slovak Research and Development Agency under contract No. APVV-0475-10 and by VEGA Scientific Grants No. 1/0592/12 and 1/0812/12 from the Ministry of Education.

References

Chaudhary, Z.I., Iqbal, J. And Rashid, J., 2003. Serum protein electrophoretic pattern in young and adult camels. Aust Vet J 81: 625-626.

Jawor, P., Steiner, S., Stefaniak, T., Baumgartner, W. And Rzasa, A., 2008. Determination of selected acute phase proteins during the treatment of limb diseases in dairy cows. Vet Med-Czech 53: 173-183.

Keay, G. and Doxey, D.I., 1982. A comparison of the serum protein electrophoretic pattern of young and adult animals. Vet Res Commun 5: 271-276.

Riond, B. Wenger-Riggenbach, B., Hofmann-Lehmann, R. and Lutz, H., 2009. Serum protein concentrations from clinically healthy horses determined by agarose gel electrophoresis. Vet Clin Pathol 38: 73-77.

Part V
Food safety quality

A mass spectrometric scoring system for oxidative damage in dairy foods

Stefan Clerens, Jeffrey E. Plowman, Stephen Haines and Jolon M. Dyer
AgResearch Ltd, Private Bag 4749, Christchurch 8140, New Zealand;
jeff.plowman@agresearch.co.nz

Introduction

Redox proteomics is an essential component of the molecular-level characterisation of food because it can provide strong evidence for protein damage resulting from processing during the preparation stage as well as the effects of packaging post-processing up to and including retail display. Since nearly all food product quality parameters correlate directly or indirectly with protein quality and function, molecular-level characterisation of protein modification and damage is of crucial importance for the food industry. For this reason it was felt that obtaining an understanding of protein damage at the amino acid residue level and correlating it with processing parameters would ultimately lead to improvements in the processing and mitigation/repair of protein damage. To this end a damage scoring system that had been developed for the characterisation of UVB-induced photo-oxidation in wool (Dyer *et al.*) was modified and refined. This refinement allowed for a more advanced evaluation of redox and non-redox protein modifications, with precise weighting and thresholding functionality using LC-MS/MS data, ensuring representative coverage across all key proteins in a sample.

Proof of concept using model proteins

Initially, in a model protein study, bovine lactoferrin and beta-lactoglobulin in phosphate buffered saline solution were irradiated with UVB light for 40 minutes to induce oxidative modifications, while control samples were left untreated. After reduction, alkylation and trypsin digestion, the resultant peptide digests were analysed by nanoLC-MS/MS. Subsequent to this the data files were converted to peak lists, imported in the ProteinScape (Bruker Daltonics, Bremen, Germany) data warehousing software, and serial Mascot searching (http://www.matrixscience.com/search_form_select.html) of the data using specific sets of variable modifications was initiated. A number of specific searches including a selection of variable modifications were conducted. Each search focused on a specific set of modifications, e.g. non-oxidative modifications such as methylation, deamidation, ubiquitylation, or on progressive sets of oxidative modifications such as found with tryptophan. The search results were then compiled per sample and exported to Excel and a series of Visual Basic scripts developed to remove redundancy, count, weight, score and sum the different damage modifications. Each targeted modification was then given a modification factor reflecting its severity, based on its rank in the modification pathway (Dyer *et al.*). The number of occurrences of each modification (multiplied by the modification factor) is then weighted against the number of observed susceptible residues. These weighted scores were then added together to determine the damage score.

In the case of the model proteins this approach resulted in a score for non-oxidative damage score (NODS), based on 6 modifications, and an oxidative damage score (ODS) based on 12 modifications. Together they added up to the total damage score (TDS).

Using this approach unmodified lactoferrin had a relatively low TDS, spilt approximately between 70% ODS and 30% NODS, whereas after UV irradiation, the lactoferrin TDS more than doubled, with oxidative damage representing approximately 90% of the TDS. In contrast lactoglobulin had a much higher base damage level compared to lactoferrin, in which oxidative damage is responsible for approximately 80% of the TDS in the control sample. After irradiation, the TDS increases by 11%, with the oxidative damage component seeing a 30% increase.

Application to commercial dairy products

To test the approach out on commercial dairy products a range of these were purchased from a local supermarket, while milk was sourced from a local farm to act as the low damage control. The whey fraction was obtained from these samples and analysed as described previously.

As expected the farm sourced milk and standard fat milk had the lowest damage levels, with slightly higher damage observed in standard fat milk powder and low fat evaporated milk (Figure 1). UHT treatment was observed to dramatically increase the TDS for this product

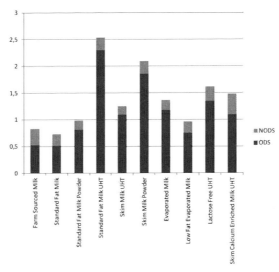

Figure 1. Damage scores obtained for nine off-the-shelf dairy products and one farm-sourced milk sample. NODS, non-oxidative damage score; ODS, oxidative damage score.

family, one UHT product giving the highest damage score in this test, underscoring the potential of UHT to induce hydrothermal damage, measured primarily as oxidative side chain modifications. Standard fat products were also observed to result in higher overall damage than low fat products, with the exception of skim milk powder, which returned a very high TDS compared to standard fat milk powder. This would be consistent with other observations from a range of proteinaceous systems in which lipid oxidation appears to have a synergistic effect with protein oxidation during damaging processing treatments or environmental insult.

References

Dyer, J.M., Plowman, J.E., Krsinic, G.L., Deb-Choudhury, S., Koehn, H., Millington, K.R. and Clerens, S., Proteomic evaluation and location of UVB-induced photo-oxidation in wool. J Photochem Photobiol B 98: 118-127.

Shotgun proteomics in blue mussels exposed to benthic trawler-induced sediment resuspension from a polluted fjord

Sara Tedesco[1], William Mullen[2], Harald Mischak[2,3], Clare Bradshaw[4] and Susana Cristobal[1,5]
[1]*Department of Clinical and Experimental Medicine, Health Science Faculty, Linköping University, Linköping, Sweden; sara.tedesco@liu.se*
[2]*Biomarker Research, Lab B 3.33a, Joseph Black Building, University of Glasgow, Glasgow, G12 8QQ, United Kingdom*
[3]*Mosaiques Diagnostics GmbH, Hannover, Germany*
[4]*Department of Systems Ecology, Stockholm University, Stockholm, Sweden*
[5]*IKERBASQUE, Basque Foundation for Science, Department of Physiology. Basque Country Medical School, Bilbao, Spain*

Objectives

Commercial bottom trawling in fishery industry is estimated to affect around 15 million km^2 of the world's seafloor every year (Bradshaw *et al.*, 2012). However, few studies have investigated whether this disturbance remobilises sediment-associated contaminants (Allan *et al.*, 2012) and, if so, whether these are bioavailable to aquatic organisms (Bradshaw *et al.*, 2012).

The blue mussel, *Mytilus edulis* is a benthic filter-feeding organism able to filter rapidly high volumes of water per day, bioaccumulating particulates and eventual pollutants (especially in digestive gland) present in the aquatic environment. The resuspension of sediments by trawling may also alter pollutants' chemical form and thus their bioavailability and toxicity (Chapman *et al.*, 1998) and/or enhance the transfer of organic pollutants in the benthic food chain, through the mobilization of contaminated particles. These repercussions in the aquatic environment can severely damage aquaculture economy and human health through fish/ seafood consumption.

Eidangerfjord, a Norwegian fjord, is characterized by sediments with high content of organic pollutants (i.e. PCDD/Fs, non-ortho PCBs and PAHs). The Norwegian authority recommend against consumption of benthic organisms from this area (Knutzen *et al.*, 2003), although a small prawn fishery weekly trawl is in this fjord (Bradshaw *et al.*, 2012).

Shotgun proteomics is basically an unexplored method in non-model aquatic organisms like blue mussels *M. edulis* and any proteomic study faced the biological effects of pollutants resuspension by bottom trawling in marine organisms.

In this context, the aim of this study was to investigate by shotgun proteomics, the protein expression in digestive gland of *M. edulis* sampled in different sites (two depths each) along this fjord subjected to continuous bottom trawling and compared with those from mussels collected in a not-trawled area much further form them. Quantitative and qualitative analysis of the differentially expressed proteins will contribute for the first time to identify specific protein expression signatures (PES) in *M. edulis* as potential biomarkers to these organic contaminants by shotgun proteomic approach in a field study.

Material and methods

The experiment was carried in June 2008 and the mussels (ca. 40 per cage, fresh from Scanfjord AB mussel farm, Lysekil, Sweden) utilized for human consumption were deployed on 3 ropes in the deep basin of Eidangerfjord (water depth 90-100 m), very close to the trawl tracks (Figure 1A) and deployed approximately 2 m and 20 m above the seabed for each rope. The mussels in the area far from trawling were only 2 m above the seabed (water depth ~80 m, Figure 1A).

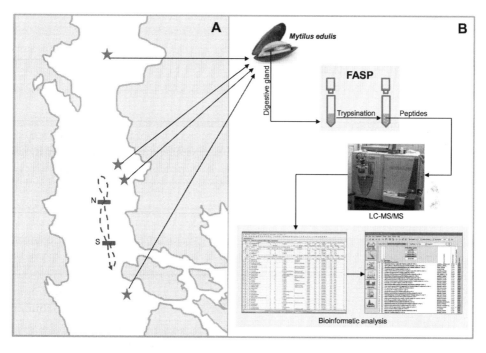

Figure 1. Schematic representation of the field experiment and methodology used in M. edulis *samples collected in Eidangerfjord. (A) Illustration of Eidangerfjord showing the position of the two paired stations for measurements of suspended sediment (squares: North and South from the previous work by (Bradshaw et al., 2012)), the path of the trawler during the experiment (dashed line) and the four sampling areas where mussels were deployed (stars). (B) A simplified diagram of the shotgun proteomic analysis performed in this work.*

Trawling by the prawn trawler Tine Marlin took place four times during the experiment (Bradshaw *et al.*, 2012) and the mussels were collected after 1 month to evaluate longer term biological effects of bottom trawling and without.

Mussels' digestive glands were therefore immediately dissected out and frozen in liquid nitrogen. Frozen samples were stored in the laboratory at -80 °C until required for analysis.

The Filter Aided Sample Preparation (FASP) technique (Wisniewski *et al.*, 2009) was utilized to extract the proteins from the digestive gland tissues prior sequence-grade trypsin digestion (Promega, UK) and liquid chromatography (Ultimate3000 system Dionex, Netherlands) coupled to LTQ Orbitrap Velos (Thermo Fisher Scientific, Germany).

The MS data were analysed by Proteome Discoverer Software (version 1.3, Thermo Scientific, Germany) matching the peptides profiles with the databases of *M. edulis*, *M. galloprovincialis* and Metazoan Phyla from UniProtKB/Swiss-Prot. After protein identification, quantitative analysis was performed by peak areas measurements after appropriate normalization (Mullen *et al.*, 2012). For a further validation of our results, the same MS data were also analysed by Scaffold software (version 3.6.2, Proteome Software Inc., USA).

A simplified representation of the methodology utilized in this work is illustrated in Figure 1B.

Results and discussion

On overage of 36 proteins per sample were identified utilizing the proteome database *M. edulis* and/or *M. galloprovincialis*, while over 1000 proteins when the taxonomy Metazoa was used as database entry. The necessity to use both databases was due to the lack of information on unsequenced proteome of *Mytilus* that reduced the amount of common proteins between the biological samples and from the different sampling areas for an exhaustive analysis valid by a statistical point of view. In addition, our findings were compared and in part confirmed by the study of contaminants found in sediments, water and in mussels collected in the same sampling areas utilized in our scientific plan (Bradshaw *et al.*, 2012).

The shotgun proteomic methodology utilized in this study on *M. edulis* required only a small amount of tissue ($\sim 3 \times 10^{-2}$ g) for each sample and more than sufficient for several technical replicates. In addition, it is fast, sensitive and reproducible technique to generate a large amount of quantitative data to carry out statistical analysis of the tissue protein composition.

The application of this proteomic approach could be advantageous not only for mytiliculture and seafood industries constrained to continuous improvement in satisfying high standards of quality, traceability and safety criteria but also a promising alternative for environmental monitoring campaigns where sample scarcity and time could affect a good and expeditious assessing of environmental pollution.

Acknowledgements

The work was funded by COST action STSM to Dr. Sara Tedesco and Dr. William Mullen. The authors would also like to thank Dr. Itxaso Apraiz and Dr. Hanna Amelina for the mussels collecting and dissection in Eidangerfjord, Dr. Angelique Stalmach and Dr. Holger Husi from University of Glasgow for their helpful advices in mass spectrometry and bioinformatics analysis.

References

Allan, I.J., Nilsson, H.C., Tjensvoll, I., Bradshaw, C. and Næs, K., 2012. PCDD/F release during benthic trawler-induced sediment resuspension. Environmental Toxicology and Chemistry 31: 2780-2787.

Bradshaw, C., Tjensvoll, I., Sköld, M., Allan, I.J., Molvaer, J., Magnusson, J., Naes, K. and Nilsson, H.C., 2012. Bottom trawling resuspends sediment and releases bioavailable contaminants in a polluted fjord. Environmental Pollution 170: 232-241.

Chapman, P.M., Wang, F., Janssen, C., Persoone, G. and Allen, H.E., 1998. Ecotoxicology of metals in aquatic sediments: binding and release, bioavailability, risk assessment, and remediation. Canadian Journal of Fisheries and Aquatic Sciences 55: 2221-2243.

Knutzen, J., Bjerkeng, B., Næs, K. and Schlabach, M., 2003. Polychlorinated dibenzofurans/dibenzo-p-dioxins (PCDF/PCDDs) and other dioxin-like substances in marine organisms from the Grenland fjords, S. Norway, 1975-2001: present contamination levels, trends and species specific accumulation of PCDF/PCDD congeners. Chemosphere 52: 745-760.

Mullen, W., Albalat, A., Gonzalez, J., Zerefos, P., Siwy, J., Franke, J. and Mischak, H., 2012. Performance of different separation methods interfaced in the same MS-reflection TOF detector: A comparison of performance between CE versus HPLC for biomarker analysis. Electrophoresis 33: 567-574.

Wisniewski, J.R., Zougman, A., Nagaraj, N. and Mann, M., 2009. Universal sample preparation method for proteome analysis. Nat Meth 6: 359-362.

Milk and cheese microbiome for safety and quality of dairy products

Alessio Soggiu[1], Emoke Bendixen[2], Milena Brasca[3], Stefano Morandi[3], Cristian Piras[1], Luigi Bonizzi[1] and Paola Roncada[4]

[1]DIVET, Dipartimento di Scienze veterinarie e sanità pubblica, Università degli Studi di Milano, Italy; alessio.soggiu@unimi.it
[2]Department of Molecular Biology and Genetics – Aarhus University, Aarhus, Denmark
[3]Institute of Sciences of Food Production, National Research Council, Milan, Italy
[4]Istituto Sperimentale Italiano L. Spallanzani, Milano, Italy

Objectives

The aim of this project is to develop and increase knowledge of key factors influencing the presence of clostridial spores in milk and factors affecting, directly or indirectly, technological characteristics and fermentation of milk, make it more or less susceptible to the development of defects (swelling). Aim of the project is to characterize by 1D/2D-gel electrophoresis and next-generation sequencing coupled to shotgun mass spectrometry the global microbiota isolated from 'Grana Padano' cheese. In particular the target is to identify the sub-proteome of *Clostridium tyrobutyricum* present in cheese and to investigate differences in proteins involved in the metabolic pathway of butyric acid in different stages of ripening and milk treatment (e.g. lysozyme treatment).

Material and methods

Clostridium tyrobutyricum strain IN15B has been isolated and identified from blowed Grana Padano Cheese samples at the Institute of Sciences of Food Production (ISPA), National Research Council (CNR), Milano, Italy as described (Cremonesi *et al.*, 2012). Pure spore suspensions of *C. tyrobutyricum* has been obtained according to the method of Yang *et al.* (Yang *et al.*, 2009).

Cheese samples were obtained from the 'Consorzio del Grana Padano' (Italy) from different factories. Samples were taken 48 h after precipitation, after 6 and 12 months of ripening. Cheeses were classified on the basis of the presence of lysozyme like an antibacterial enzyme, e.g. those with lysozyme (MN and PC) and those without lysozyme (TN). Another characteristic analysed was the presence (bad) or absence (good) of holes or blowing phenomena in cheese. Aqueous phases were extracted from cheese aliquots after 12 months of ripening according to (Jardin *et al.*, 2012) with some modification: 300 mg grated Grana Padano cheese were dispersed in 1ml of MilliQ water (Millipore, Germany) and homogenized with ultrasounds (Bandelin SONOPULS, Germany) for 2 min at 50% power and further stirred for 1 h at 40 °C.

Cheese samples were centrifuged at 10,000×g for 10 min at 20 °C, to separate the aqueous phase from the pelleted caseins.

Aqueous phases from cheese were concentrated about 10 times on a speed-vac apparatus (Thermo Scientific, Denmark), mixed with cold acetone (1:6 v/v) and precipitated at -20 °C overnight. Protein pellets were re-suspended in denaturing buffer (8 M urea in 100 mM ammonium bicarbonate for TRIPLE-TOF analysis). Proteins were quantitated using a Bradford assay (Biorad, USA).

Bacterial protein extraction was performed as previously described (Piras *et al.*, 2012). Cellular pellets were re-suspended in denaturing buffer (8 M urea in 100 mM ammonium bicarbonate). Protein concentration in all samples was determined using a Bradford kit (Biorad, USA). Bacterial proteins were reduced, alkylated and digested. The obtained peptides purificated using Poros R2 resin. LC–MS/MS analyses were performed on an EASY-nLC II system (Thermo Scientific, Denmark) connected to a TripleTOF 5600 mass spectrometer (AB Sciex, Denmark). Cheese protein extracts were digested and subjected to SCX fractionation and LC-MS/MS with a Q-Star Elite (Applied Biosystems, USA) as described in Nissen *et al.* (2012). The collected MS files were converted to Mascot generic format (MGF). The peak lists were used to interrogate several databases using Mascot 2.3.02 (Matrix Science,UK). The significance threshold (*P*) was set at 0.01

Results and discussion

Bacterial protein identification

Preliminary shotgun data from pure protein suspensions of *C. tyrobutyricum* bacteria and spore were obtained using a AB5600TripleTOF mass spectrometer. Mascot generic files (MGF) were searched against the UniprotKB database, the NCBInr database and the cheese custom database as described in the methods paragraph. Table 1 reports the total number of bacterial proteins identified from each database

Cheese metaproteomics

Preliminary shotgun data from casein depleted cheese samples were obtained using a Q-Star Elite mass spectrometer. Mascot generic files (MGF) were searched against the uniprotKB database, the NCBInr database and the cheese custom database as described in the methods paragraph. Table 2 reports the total number of proteins identified from each database.

In this work proteins both from bacterial and bovine (also chicken-egg lysozyme) origin were identified in the cheese matrix containing or not lysozyme (an antibacterial addictive) by a shotgun approach. The identification of bacterial proteins released into the cheese aqueous

Table 1. Mascot data statistics about bacterial proteins identified by TripleTOF shotgun analysis.

Mascot database	taxa	*C. tyrobutyricum* bacterial proteins (unassigned peptides)	FDR%	*C. tyrobutyricum* spore proteins (unassigned peptides)	FDR%
Swissprot	Taxa	88 (7,394)	2.59	53 (8,021)	12.50
NCBInr	bacteria	133 (7,175)	3.09	67 (7,940)	3.75
Swissprot	Taxa	10 (7,837)	74.07	5 (8,178)	50
NCBInr	mammalia	7 (7,842)	76.67	6 (8,144)	17.74
CHEESE	All	73 (7,335)	2.81	32 (7,904)	1.37

Significance threshold $P<0.01$ for all protein identified.

Table 2. Mascot data statistics about all cheese proteins identified by Q-TOF shotgun analysis.

Mascot database	taxa	MN427 12 months good 001 (unassigned peptides)	FDR%	MN427 12 months blowed 004 (unassigned peptides)	FDR%	TN305 12 months good 002 (unassigned peptides)	FDR%	TN305 12 months blowed 003 (unassigned peptides)	FDR%
Swissprot	Taxa	6	23.08	8	12.50	9	38	22	13
NCBInr	bacteria	3				10	18.18		
Swissprot	Taxa	13		13	1.21	15	2.79	38	1.17
NCBInr	mammalia	17		15	7.69	17	1.30		
CHEESE	All	20 (3,563)	1.55	15 (2,593)	1.82	28 (4,425)	2.08	68 (14,141)	1.61

Significance threshold $P<0.01$ for all protein identified.

phase was very difficult because of the great dynamic range of protein concentration between milk and bacterial proteins.

To decrease this dynamic range we simplified the proteome before MS analysis through a pre-fractionation step, which included a centrifugation step to remove caseins from aqueous phase. However, complexity remained and has influenced the identification of bacterial low abundance proteins. Other reasons for the low number of proteins identified could be: low protein abundance and the co-elution of peptides with other highly abundant peptides. Otherwise the main problem in this study remains the lack of annotated bacterial protein sequences in all the databases used. In fact the genome of *C. tyrobutyricum*, the main contaminant in Grana Padano cheese, at this moment remains unsequenced.

MS data obtained until now show a large number of unmatched peptides (>10,000) that hopefully we will identify in the future using databases obtained from next-generation sequencing data. Meanwhile we tried to use a custom database built as described in methods section. Preliminary results show a good number of clostridial bacterial protein identified without redundancies and with a good FDR compared to the other database used.

We expect that our building of metaproteomic information of Grana Padano cheese will enhance safety and quality of this product. And also shed light to understand the relationship between metabolism of starter and contaminant bacteria and cheese composition.

Acknowledgements

This mission was supported by a travel grant from the European Cooperation in Science and Technology (COST) for a short term scientific mission (STSM) from the COST action FA1002. The author is supported by a post-doc grant from UNIMI-DIVET department and from FILIGRANA project, financed from MiPAAF D.M 25741/7303/11 – 01/12/2011. I am sincerely grateful to Pr. Bendixen for her strong scientific and human support during this month, and to her strong encouragement and her precious advices about this research project. I would like to thank also all the group of Pr. Enghild to host me during experiments.

References

Cremonesi, P., Vanoni, L., Silvetti, T., Morandi, S. and Brasca, M., 2012. Identification of *Clostridium beijerinckii*, *Cl. butyricum, Cl. sporogenes, Cl. tyrobutyricum* isolated from silage, raw milk and hard cheese by a multiplex PCR assay. J Dairy Res 79: 318-323.

Jardin, J., Molle, D., Piot, M., Lortal, S. and Gagnaire, V., 2012. Quantitative proteomic analysis of bacterial enzymes released in cheese during ripening. Int J Food Microbiol 155: 19-28.

Nissen, A., Bendixen, E., Ingvartsen, K.L. and Rontved, C.M., 2012. In-depth analysis of low abundant proteins in bovine colostrum using different fractionation techniques. Proteomics 12: 2866-2878.

Piras, C., Soggiu, A., Bonizzi, L., Gaviraghi, A., Deriu, F., De Martino, L., Iovane, G., Amoresano, A. and Roncada, P., 2012. Comparative proteomics to evaluate multi drug resistance in *Escherichia coli*. Mol Biosyst 8: 1060-1067.

Yang, W.W., Crow-Willard, E.N. and Ponce, A., 2009. Production and characterization of pure *Clostridium* spore suspensions. J Appl Microbiol 106: 27-33.

Molecular characterization of Maltese honey: diastase and proline levels changes in Maltese honey seasons

Adrian Bugeja Douglas, Everaldo Attard and Charles Camilleri
University of Malta, Institute of Earth Systems, Division of Rural Sciences and Food Systems,
Malta; adrian.bugeja-douglas@um.edu.mt

Objectives

A research project between the Division of Rural Sciences and Food Systems at the University of Malta and Golden Island Ltd. Is currently analysing in detail the physicochemical characteristics of Maltese Honey following internationally recognized standard techniques. Among the parameters being analysed are the amino acid proline and the enzyme diastase.

History of Maltese honey

The Maltese Islands have been renowned for the production of high quality honey, since ancient times. The Greeks called the island Μελίτη (Melite) meaning 'honey-sweet', for the unique production of high quality honey. The name Melite was also used under the Roman rule. It was under the Arab rule that the name of Melite was changed to Malta.

Today after millennia, Maltese honey still remains a well sought and prized gourmet product. The Maltese Island has three honey seasons; each honey season has its own characteristic organoleptic characteristics:
1. Spring honey is a multi-floral honey;
2. Summer honey is manly based on thyme;
3. Autumn honey is based on Eucalyptus and Carob.

Characteristics of honey

Honey is composed mainly from carbohydrates, a small percentage of water and a vast array of other substances. These other substances, which are found in trace amounts include: amino acids, proteins, enzymes, vitamins, minerals, organic acids, and solid particles.

Carbohydrate content comprises about 95% of honey dry weight. Main sugars are the monosaccharides hexoses; fructose (about 38.5%) and glucose (about 31.0%), which are products of the hydrolysis of the disaccharide sucrose.

Honey has a relatively low content of acid, but the acid plays an important role to the taste of honey. The acids are added by the bees and the main acid is gluconic acid. The pH range of honey varies from 3.3 to 4.6.

The amounts of amino acids and proteins are quite small (about 0.7%); however they are important for judging the honey quality. Honey contains the most physiological important amino acids (Perez *et al.*, 2007). The amino acid proline, which is added by bees, is a measure of honey ripeness. The proline content of normal honeys should be more than 200mg/kg. Values below 180mg/kg mean that the honey is probably adulterated. The other amino acids do not play a key role for the determination of quality or origin of honey.

The honey proteins are mainly enzymes, which are added by the bees during the process of honey ripening. Diastase (amylase) digests starch to maltose. Invertase converts sucrose to glucose and fructose. Diastase plays an important role for judging the honey quality and is used as indicators of honey freshness. A minimum value of 10 diastase units is set in the European Honey Directive. Their activity decays upon storage and if the honey is exposed to heat. Also the floral origin of honey is known to influence the amounts of diastase

Material and methods

For each season (spring, summer and autumn) 30 samples of honey were collected from beekeepers. As far as possible we tried to collect samples so that the whole geographical area of the Maltese Islands (Malta and Gozo). Samples were collected in 150 mls jars, labelled and stored in a fridge (4 °C) until day of analysis so as to conserve the amino acids and protein content.

Proline

Definition: The content of proline is defined as the colour developed with ninhydrin, compared with a proline standard and expressed as a proportion of the mass of honey in mg/kg. The method is based on the original method of Ough (Ough, 1969) and as adapted by the International Honey Commission in Harmonised Methods of the International Honey Commission (Bogdanov, 2009).

Sample preparation: The jars containing the honey were inverted several times to produce a homogenized sample. 5.00 g of honey were weighed to the nearest mg in a 100 ml beaker and dissolved in 50 ml of distilled water. The honey solutions were quantitatively transferred to a 100 ml volumetric flask and diluted to volume with water and shaken well.

Determination of proline level: The coefficient of extinction is not constant. Therefore, for each series of measurements the average of the extinction coefficient of the proline standard solution was determined in triplicate.

In 15 ml centrifuge tubes, the following solutions were set-up: 0.5 ml of the sample solution in one tube, 0.5 ml of water (blank test) into a second tube and 0.5 ml of proline standard solution into three other tubes. To each tube, 1 ml of formic acid and 1 ml of ninhydrin solution (3% of ninhydrin in ethylene glycol monomethylether) were added. Each tube was caped and vigorously shaken for 15 minutes in a shaker. The tubes were then placed in a water bath at 100 °C for 15 minutes. The tubes were immersed below the level of the solution. The tubes were then transferred to a water bath at 70 °C for 10 minutes. To each tube, 5 ml of 2-propanol-water-solution (1:1 solution of 2-propanol and distilled water) was added and tubes were caped immediately. Afterwards the tubes were left to cool for 45 min. The absorbance of each solution was measured with a spectrophotometer at a scan from 5,000 nm to 540 nm, using 1 cm cells

Calculation

Proline in mg/kg honey at one decimal place was calculated according to following equation:

$$\text{Proline (mg/kg)} = \frac{Es}{Ea} \times \frac{E_1}{E_2} \times 80$$

where
Es = absorbance of the sample solution;
Ea = absorbance of the proline standard solution (average of three readings);
E_1 = mg proline taken for the standard solution;
E_2 = weight of honey in grams;
80 = dilution factor.

Diastase

Definition: The unit of Diastase Activity, (Schade), is defined as that amount of enzyme which will convert 0.01 grams of starch to the prescribed end point in an hour at 40 °C under the conditions of the test. The method is based on the Alpha-Amylase Test by Megazyme, Ireland (T-AMZ200).

Sample preparation: The jars containing the honey were inverted several times to produce a homogenized sample. 2.00 g of honey were weighed into a 100 ml beaker and dissolved in 40 ml of 100 mM sodium maleate buffer (pH 5.6). The sample solutions were quantitatively transferred to 50 ml volumetric flask and adjusted to volume.

Determination of proline level: 1.0 ml of diluted honey samples were dispensed to the bottom of 16×120 mm tubes and pre-incubate at 40 °C for 5 min in a water bath. Using a forceps, a tablet of Amylazyme tablet (Megazyme Ireland, lot number 30602) was added to the each tube. The tubes were incubated at 40 °C for exactly 10 min in the water bath. 10 ml of Trizma base solution (2% w/v, Sigma cat. no. T-1503) were added to terminate the reactions. The tubes were then stirred vigorously on a vortex mixer. Afterwards each tube was placed in a stand and allowed to sit at room temperature. After approximately 5 min, the tubes were stirred again and the solutions were filtered through a Whatman No.1 (9 cm) filter paper. The absorbance of the sample solutions were measured at 590 nm against a reaction blank (solution lacking honey sample), using a spectrophotometer.

Calculation of activity

The α-amylase activity of a sample (as Schade units per gram of honey) was determined by using the associated regression equation.

Diastase activity (DN) (Schade units/gram of honey)
$$= 21.2 \times \Delta590 \text{ (Amylazyme Lot 30602)} + 0.27 - 2.9 \times \Delta590^2$$

If an absorbance value for a sample was below 0.30, the sample was re-assayed and the incubation time increased to 20 min (doubled). Diastatic activity calculated from the regression equation was then divided by 2 to give the correct values in Schade units/gram.

Results and discussion

From the results obtained, we observed that there are no seasonal variations for the diastase results (1.092-19.619 Schade Units). On the other hand, for the proline values we observed that those for spring and summer (x=0.5909 and 0.5789 g/kg) were significantly different from the autumn (x=0.7367 g/kg) results ($P<0.05$, v=46). There was no correlation between the diastase and proline values.

Acknowledgements

Financial support was from The Malta Council for Science & Technology through the National Research and Innovation Programme 2010.

References

Bogdanov, S., 2009. Harmoised methods of the Internation Honey commision. http://www.bee-hexagon.net/en/network.htm.

Ough, C., 1969. Rapid determination of proline in grapes and wines. J. Food Science 34: 228-230.

Perez, R.A., Iglesias, M.T., Pueyo, E., Gonzalez, M., De Lorenzo, C., 2007. Amino acid composition and antioxidant capacity of Spanish honeys. Journal of agricultural and food chemistry 55: 360-365.

Untargeted metabolomic analyses open new scenarios in *post mortem* pig muscles: Casertana and Large White

Cristina Marrocco, Angelo D'Alessandro, Sara Rinalducci, Cristiana Mirasole and Lello Zolla
Laboratory of Proteomics and Mass Spectrometry, Department of Ecological and Biological Sciences, Tuscia University, Viterbo, Italy; zolla@unitus.it

Objectives

Post mortem metabolism of pig muscles represents a pivotal area of research, in that pig meat quality is largely influenced by biochemical changes arising soon after slaughter. Metabolic rate and, in particular, forcedly anaerobic glycolytic fluxes in *post mortem* muscles are indeed correlated with the final meat quality outcome, especially in the light of the negative influence of a rapid pH drop in the determination of the so-called Pale Soft Exudative phenotype.

Centuries of livestock breeding selection have promoted the diffusion of the fast growing lean meat producing Large White pigs, while local breeds characterized by slower growth rates and higher backfat and intramuscular accumulation have been progressively sacrificed (especially in western countries) on the altar of productivity, often underestimating the quality (taste, flavor) issue. *Longissimus lumborum* muscles from the local high-fat depositing and flavored meat-producing Casertana pig breed were investigated through proteomics (Murgiano *et al.*, 2010) and Integrated Omics (D'Alessandro *et al.*, 2011a) in comparison to Large White counterparts. Meat quality parameters (early and ultimate post-mortem pH, water holding capacity and Minolta L×a×b*values) were also assayed and correlated with relative abundances of specific proteins and metabolites, by exploiting supervised (*ad hoc*) approaches, such as targeted metabolomics. Proteomics and targeted metabolomics investigations confirmed a solid correlation between *post mortem* metabolic rates (including both energy metabolism and redox poise modulation). However, the intrinsic limitation of the targeted metabolomic investigation that we had previously proposed (D'Alessandro *et al.*, 2011a) did not lend itself to fully cover the whole metabolome. This is a significant pitfall of targeted approaches, in that they do not lend themselves to extensively complement proteomics results and solve some intricacies arising from proteomics analyses, such as those related to relative quantities of glycolytic enzymes showing increased levels in Casertana (phosphoglucomutase, glyceraldehyde 3-phosphate dehydrogenase, enolase, lactate dehydrogenase), and a few, albeit rate-limiting (pyruvate kinase), in Large White. Indeed, targeted approaches are not suitable for exploratory studies, in that they are limited to the assessment of a subset of metabolites through aimed analytical strategies, such as Multiple Reaction Monitoring (D'Alessandro *et al.*, 2011b).

In this view, we hereby applied an optimized HPLC-Q-TOF mass spectrometry-based approach for untargeted metabolomics analyes of Casertana and Large White *longissimus lumborum* muscles. Analyses of the mass spectra was eased by the application of a recent freely available software suite that enables in depth analyses of accurate MS spectra (MAVEN – Melamud *et al.*, 2010).

Material and methods

Animal handling followed the recommendations of European Union directive 86/609/EEC and Italian law 116/92 concerning animal care. In a commercial dairy farm located in Viterbo (Italy) and Campobasso (Italy), we selected 30 animals (15 per breed; live weight of 207.80 + 8.53 and 140.08 + 10.58 for Large White and Casertana, respectively) which were of the same age and fed the same diet. Samples from each Casertana and Large White individuals (150 mg of *longissimus lumborum* muscles at 24 h after slaughter) were extracted as previously reported (D'Alessandro *et al.*, 2011a). The extracted samples were re-suspended in 1 ml of water, 5% formic acid and transferred to glass autosampler vials for LC/MS analysis.

An Ultimate 3000 Rapid Resolution HPLC system (LC Packings, DIONEX, Sunnyvale, USA) was used to perform metabolite separation. The system featured a binary pump and vacuum degasser, well-plate autosampler with a six-port micro-switching valve, a thermostated column compartment. Samples were loaded onto a Reprosil C18 column (2.0×150 mm, 2.5 µm – Dr Maisch, Germany) for metabolite separation. Chromatographic separations were achieved at a column temperature of 30 °C; and flow rate of 0.2 ml/min. For downstream positive ion mode (+) MS analyses, a 0-100% linear gradient of solvent A (ddH2O, 0.1% formic acid) to B (acetonitrile, 0.1% formic acid) was employed over 30 min, returning to 100% A in 2 minutes and a 6-min post-time solvent A hold. Due to the use of linear ion counting for direct comparisons against naturally expected isotopic ratios, time-of-flight instruments are most often the best choice for molecular formula determination. Thus, mass spectrometry analysis was carried out on an electrospray hybrid quadrupole time-of flight mass spectrometer MicroTOF-Q (Bruker-Daltonik, Bremen, Germany) equipped with an ESI-ion source. Mass spectra for metabolite extracted samples were acquired both in positive and in negative ion mode. ESI capillary voltage was set at 4,500 V (+) ion mode. The liquid nebulizer was set to 27 psi and the nitrogen drying gas was set to a flow rate of 6 l/min. Dry gas temperature was maintained at 200 °C. Data were acquired with a stored mass range of m/z 50-1,200. Calibration of the mass analyzer is essential in order to maintain an high level of mass accuracy. Instrument calibration was performed externally every day with a sodium formate solution consisting of 10 mM sodium hydroxide in 50% isopropanol: water, 0.1% formic acid. Automated internal mass scale calibration was performed through direct automated injection of the calibration solution at the beginning and at the end of each run by a 6-port divert-valve.

Mass spectrometry chromatograms were elaborated for peak alignment, matching and comparison of parent and fragment ions, and tentative metabolite identification (within a

20 ppm mass-deviation range between observed and expected results against the KEGG database – Kanehisa and Goto, 2000). Quantitative fluctuations were plotted as node size variations in Large White against Casertana values.

Results and discussions

Untargeted metabolomics analyses allow unprecedented monitoring of metabolic species in swine muscles after slaughter. An overview of the whole metabolome (KEGG pathway no: ko01100) is provided in Figure 1, upon normalization of Large White values against Casertana counterparts. Through this approach, we could confirm previous MRM-based

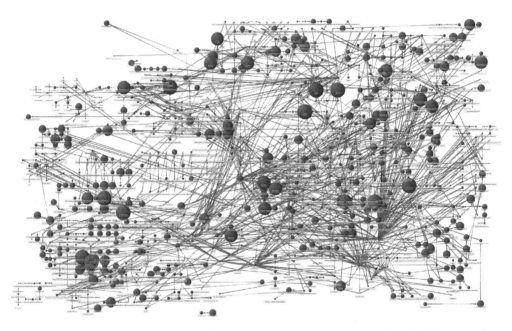

Figure 1. Overview of the whole metabolome (KEGG pathway no: ko01100) of Large White longissimus lumborum, *normalized against Casertana counterparts. The metabolic pathway has been obtained from the KEGG pathway database (Ko01100 – Kanehisa and Goto, 2000) and re-elaborated with the output from untargeted metabolomics analysis, upon automatic best peak selection at 20 ppm error through the MAVEN software package (Melamud et al., 2010). Each node represents a distinct metabolite, whose relative intensity has been normalized to the relative intensity values detected in Casertana. Directed edges indicate metabolic fluxes within each graphed pathway. The immediacy of the interpretation of these maps stems from the possibility to determine the relative activation/suppression of a specific metabolic cycle at a rapid glance. Indeed, node sizes are mostly altered throughout the whole metabolome in Large White pigs, which is consistent with the increased metabolic rate in fast growing Large White pigs in comparison to Casertana.*

results (higher levels of lactate, glycerol 3-phosphate and phosphocreatine in Casertana, higher GSH levels and lower GSSG in Large White), while we could also complement the previously collected information with a whole new array of metabolites, including those involved in fatty acid biosynthesis and metabolism, Kreb's cycle, pentose phosphate pathway, aminoacid biosynthesis, arginine and proline metabolism. Among those results which deserve further in-depth investigations, we also observed an increased rate of the pentose phosphate pathway (especially at non-oxidative steps) in Large White pigs, which is consistent with the anabolic requirements for NADPH in this fast growing pig breed.

Acknowledgments

This study has been supported by the 'GENZOOT' research programme, funded by the Italian Ministry of Agricultural, Food and Forestry Policies (Ministero delle Politiche Agricole, Alimentari e Forestali).

The 'Nutrigenomica Mediterranea: dalla nutrizione molecolare alla valorizzazione dei prodotti tipici della dieta mediterranea – NUME' project, funded by the Italian Ministry of Agricultural, Food and Forestry Policies (Ministero delle Politiche Agricole, Alimentari e Forestali).

References

D'Alessandro, A., Gevi, F., Zolla, L., 2011b. A robust high resolution reversed-phase HPLC strategy to investigate various metabolic species in different biological models. Mol Biosyst. 7(4):1024-1032.

D'Alessandro, A., Marrocco, C., Zolla, V., D'Andrea, M., Zolla, L., 2011a. Meat quality of the *longissimus lumborum* muscle of Casertana and Large White pigs: metabolomics and proteomics intertwined. J Proteomics. 75(2):610-627.

Kanehisa, M., Goto, S., 2000. KEGG: kyoto encyclopedia of genes and genomes. Nucleic Acids Res. 28(1):27-30.

Melamud, E., Vastag L., Rabinowitz J.D., 2010. Metabolomic analysis and visualization engine for LC-MS data, Anal. Chem. 82(23), 9818-9826.

Murgiano, L., D'Alessandro, A., Egidi, M.G., Crisà, A., Prosperini, G., Timperio, A.M., Valentini, A., Zolla, L., 2010. Proteomics and transcriptomics investigation on longissimus muscles in Large White and Casertana pig breeds. J Proteome Res. 9(12):6450-6466.

The role of salt in dry cured ham processing characterized by LC-MS/MS-based proteomics

Gianluca Paredi[1], Samanta Raboni[1],Adam Dowle[2], David Ashford[2], Jerry Thomas[2], Jane Thomas-Oates[3],Giovanna Saccani[4], Roberta Virgili[4] and Andrea Mozzarelli[1,5]
[1]Department of Pharmacy, Interdepartmental Center Siteia. Parma, University of Parma, Parma, Italy; gianluca.paredi@unipr.it
[2]Centre of Excellence in Mass Spectrometry and Technology Facility, Department of Biology, University of York, York, United Kingdom
[3]Centre of Excellence in Mass Spectrometry and Department of Chemistry, University of York, York, United Kingdom
[4]Stazione Sperimentale per l'Industria delle Conserve Alimentari, Parma, Italy
[5]National Institute of Biostructures and Biosystems, Rome, Italy

Introduction

Parma dry-cured ham is a typical Italian food product, manufactured according to a technological process based on empirical observations and traditional recipes, consisting of steps of salting, resting, drying and ageing. The salting phase is crucial because salt acts as a bacteriostatic agent, reduces water content, and plays an active role in the development of the sensory and quality parameters of the final product. The salting consists of two phases: in the first phase, the pork legs are covered with sea salt and then left in a room for 5-7 days at controlled temperature (0-3 °C) and humidity (75-90%). In the second phase, the hams, after the removal of the remained salt, are covered again with fresh sea salt and stored for 12-15 days in the above mentioned environmental conditions. Two kinds of salt are used to cover the pork legs. A wet salt is used in the ham rind while dry salt is used to cover the unskinned parts. During the salting phases an exudate rich in proteins is produced. We have determined by LC-MS/MS the protein pattern of exudates obtained under different technological conditions.

Material and methods

Eight fresh legs from domestic heavy pigs, average weight 13.20±0.22 kg, were selected for the analysis. Four legs were pressed (12 kg/ham for 24 h) perpendicular to the fat-free muscle mask. The pressure step was carried out to obtain a decrease in both ham thickness and shape variability. The standard shape was preserved for the other four legs. During the salting phase, exudates were collected from each leg after 1, 5 and 18 days. The exudates were centrifuged at 16,000×g for 15 minutes in order to remove insoluble material. Due to the high salt concentration, 1 ml of each sample solution was dialyzed against 40 mMTris, 0.5% SDS, pH 7.4, for 24 hours. Protein concentration was determined using the Bradford assay with bovine serum albumin for calibration.

The samples were processed using two protocols: (1) 200 μg of sample were processedessentially according to the FASP procedure reported by Wiśniewski *(Wisniewski et al., 2009)* using filter units with a nominal cut-off of 30 kDa; (2) 20 μg of sample were separated using monodimensional electrophoresis for 5 minutes at 200 V using NuPAGE Novex Bis-Tris gels 10% T% (Invitrogen). The resultant single broad bands were excised and digested in-gel. The concentration of the resulting peptide solution was estimated by nanodrop UV spectrophotometry. LC-MS/MS analysis was carried out using a nano Acquity UPLC system (Waters) coupled to a maXis LC-MS/MS System (BrukerDaltonics) with a nano-electrospray source. Chromatographic separation of peptides was performed on a C_{18} reversed phase column using either a short (20 min) or a long (125 min) gradient of 5-35% acetonitrile in aqueous0.1% (v/v) formic acid. Positive ESI-MS and MS/MS spectra were acquired using Auto MSMS mode. The resulting MS/MS spectra were submitted to database searching using the MASCOT search engine against the Swiss Protdatabase (31/10/2012). The search was restricted to mammalian proteins, and two missed cleavages were allowed. Carbamidomethyl modification was set as a fixed modification, whereas methionine oxidation was set as a variable modification. A value of 10 ppm was used for the peptide tolerance and a value of 0.1 Da for the fragment tolerance. A significance threshold and expect score cut-off of 0.05 were applied.

Results and discussion

Aninitial evaluation of the exudates, processed using the FASP or gel protocols, was carried out applying the short HPLC gradient. An average of 55.2 and 41.8 proteins were identified in the FASP and gel samples, respectively. The highest number of identification, 63 proteins, was obtained for the sample of unpressed meat recovered on day 18, while the lowest was obtained for the day 1sample of unpressed meat. 42 proteins were found to be common to the FASP and gel protocols while 20 and 24 proteins were unique to gel and FASP samples, respectively. Due to the higher number of proteins identified, the FASP samples were processed with the long gradient protocol and with quantitative emPAI and gene ontology (GO) analysis. Overall, 260 proteins were identified in the six samples. Only 15% of them belong to the extracellular region while most of them belong to the cytosolic component (Figure 1A). Hence, salt penetration into the legs in the salting phase is concluded to be a harsh process that leads to cell rupture and the extraction of intracellular proteins. The overall molecular weight distribution analysis (Figure 1 B) showed that 43% of proteins exhibit a molecular weight under 40 kDa and, from the isoelectric point distribution (Figure 1C), 55% of proteins exhibit a pI between 5 and 7. EmPAI analysis (Ishihama *et al.*, 2005) was carried out comparing the pressed and unpressed samples, at different sampling days. For day 1 samples, 91 proteins were common to the two conditions and 28 showed almost a two-fold variation. Specifically, 20 proteins were more concentrated in the pressed sample. Interestingly, several myofibrillar proteins, like tropomyosin 1 and 2, troponin C, troponin I, troponin T, were more concentrated in the pressed sample. As showed in previous studies, these proteins undergo to massive proteolysis degradation during the ripening phase, strongly influencing the flavor and quality of the final product (Kitamura *et al.*, 2006; Luccia *et al.*, 2005; Monin *et al.*, 1997; Toldra *et al.*, 1993). The higher quantity in the exudates of pressed meat might be due to alower proteolysis activity caused by a higher concentration of salt. For day 5 samples, 16 proteins showed a change in quantity, with 5 being more concentrated in pressed samples and 11 in the unpressed samples. Finally, a different behavior was observed for 30 proteins in the day 18 samples, with 10 more concentrated in the pressed samples and 20 in the unpressed samples. In this group, myoglobin wasthree foldmore concentrated in pressed samples than in unpressed samples. Considering the change as a function of time, 23 proteins showed a varying extraction profile in the three pressed samples. Specifically, seven proteins were more concentrated in day 1 pressed samples, one in day 5 pressed samples and three in day 18 pressed samples. In contrast, the level of six proteins was lower in day 1 pressed samples, three in day 5 pressed samples and three in day 18 pressed samples. In the unpressed samples, the level of 29 proteins changed in the three samples. For day 1 unpressed samples, four proteins were more concentrated and nine less concentrated in comparison with day 5 and day 18 samples. For day 5 unpressed samples, three proteins were more concentrated and three less concentrated than day 1 and day 18. Finally, seven proteins were preferentially extracted in day 18 unpressed samples. The extraction of the myofibrillar proteins tropomyosin 1 and 2 was constantly increased from day 1 to 18 in

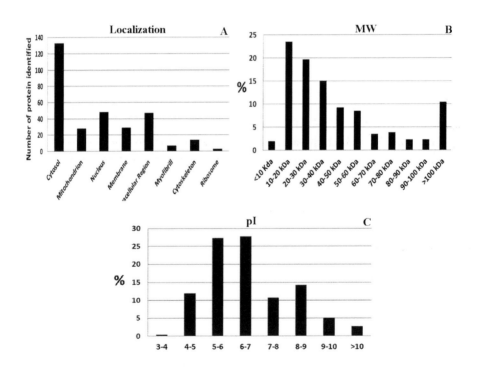

Figure 1. Histograms of main localization, MW and pI. Main predicted localization (A), molecular weight distribution (B) and isoelectric point distribution (C) of the proteins identified with the long gradient protocol.

both pressed and unpressed samples, whereas troponin C, troponin I and troponin T showed the same behavior only in unpressed samples.

Conclusions

We have found that the pressing step highly influences the extraction of proteins during the salting phase. Particularly, changes in the extraction profile of several myofibrillar proteins, that are associated to quality of dry cured ham, were detected. These results clearly shows that the proteomics analysis is suitable (1) to highlight differences triggered by distinct technological parameters, and (2) to direct further process modifications toward the preparation of high quality dry-cured ham at low salt concentration, a stringent requirement for consumer health.

Acknowledgements

This work was in part carried out with the financial support provided by COST action FA1002-Proteomics in Farm Animals to Gianluca Paredi with a STSMs fellowship at laboratory of Jane Thomas-Oates.

References

Ishihama, Y., Oda, Y., Tabata, T., Sato, T., Nagasu, T., Rappsilber, J. and Mann, M., 2005. Exponentially modified protein abundance index (emPAI) for estimation of absolute protein amount in proteomics by the number of sequenced peptides per protein. Mol Cell Proteomics 4: 1265-1272.

Kitamura, S., Muroya, S., Nakajima, I., Chikuni, K. and Nishimura, T., 2006. Amino acid sequences of porcine fast and slow troponin T isoforms. Biosci Biotechnol Biochem 70: 726-728.

Luccia, A.D., Picariello, G., Cacace, G., Scaloni, A., Faccia, M., Liuzzi, V., Alviti, G. and Musso, S.S., 2005. Proteomic analysis of water soluble and myofibrillar protein changes occurring in dry-cured hams. Meat Sci 69: 479-491.

Monin, G., Marinova, P., Talmant, A., Martin, J.F., Cornet, M., Lanore, D. and Grasso, F., 1997. Chemical and structural changes in dry-cured hams (Bayonne hams) during processing and effects of the dehairing technique. Meat Sci 47: 29-47.

Toldra, F., Cervero, M.C. and Part, C., 1993. Porcine Aminopeptidase Activity as Affected by Curing Agents. Journal of Food Science 58: 724-&.

Wisniewski, J.R., Zougman, A., Nagaraj, N. and Mann, M., 2009. Universal sample preparation method for proteome analysis. Nat Methods 6: 359-362.

Proteolytic action of caspases 3 and 7 on the hydrolysis of bovine and porcine muscle myofibrillar proteins

Rosa A. Rodríguez-Frómeta[1], Jesús Rodríguez-Díaz[1], Enrique Sentandreu[1], Ahmed Ouali[2], Miguel A. Sentandreu[1]
[1]*Instituto de Agroquímica y Tecnología de Alimentos (CSIC), Avenida Agustín Escardino, 7, 46980 Paterna (Valencia), Spain; ciesen@iata.csic.es*
[2]*UR370, QuaPA, INRA de Clermont Ferrand Theix, St. Genès Champanelle, France*

Introduction

The conversion of muscle into meat occurs immediately after animal slaughter and implies numerous biochemical reactions that nowadays are far to be fully understood. Research carried out during the last years has demonstrated that one of the first events occurring in postmortem muscle is the onset of cellular death or apoptosis (Ouali *et al.* 2006). This process is mediated by a particular group of enzymes called caspases. Triggering of this phenomenon would have important consequences for the rest of other better known processes occurring postmortem, such as the establishment of rigor mortis due to ATP depletion and the breakdown of the myofibrillar structure due to the action of several proteolytic enzyme groups. The final consequence will be the reduction of the mechanical resistance of meat, thus giving rise to tender meat (D'Alessandro *et al.* 2012). One of the main problems that meat industry has to face is inconsistency of the final meat tenderness, together with the impossibility to explain and predict this variability. In an effort to better understand the postmortem proteolysis of myofibrillar proteins by muscle peptidases and its relation to the development of meat tenderness, the present work had the objective to evaluate the potential action of executioner caspases 3 and 7 on the degradation of different myofibrillar proteins.

Material and methods

Production of recombinant caspases 3 and 7

Addgene plasmids pET23b-Casp3-His and pET23b-Casp7-His (a gift of Dr. Salvesen) were transformed into competent cells of *E. coli* BL21 pLyS. The cells were grown in LB containing ampicilline and cloramphenicol to an OD_{500} of 0.6. Cells were then induced with 0.2 mM IPTG and incubated for 4 h at 25 °C 200 rpm. After induction cells were harvested by centrifugation at 7,000 rpm during 20 min (4 °C). The pellet containing the cells was resuspended in lysis buffer (Tris 100 mM pH8, NaCl 100 mM) and freeze thawed to produce the cellular lysis. After centrifugation at 15,000 rpm for 30 min, the supernatant was used as the source material for the purification of recombinant caspases 3 and 7 on Ni-NTA-Agarose. Elution of proteins was done by applying a linear gradient of imidazol. Caspase 3 eluted from the column at 300 mM imidazol, whereas caspase 7 eluted considerably later in the gradient at 600 mM imidazol. Collected fractions containing caspases were desalted and concentrated using centrifugal filters prior to use in the incubation studies.

Preparation of myofibrilar protein extracts

One gram of either bovine or porcine raw meat samples was homogenized in 10 ml of 50 mM Tris buffer, pH 8.0. The homogenate was then centrifuged at 10,000 g for 20 min at 4 °C, collecting the precipitate that was further redissolved it in Tris buffer, pH 8.0, containing 6 M urea and 1 M thiourea.

Incubation of myofibrillar extracts with recombinant caspases 3 and 7 and analysis by SDS-PAGE

The obtained bovine and porcine protein extracts were diluted to a protein concentration of 1 mg/ml, then taking 400 µl of each extract for incubation with 100 µl of a purified preparation of recombinant caspase 3 or 7. The mixture was incubated at room temperature for 48 h, inactivating the caspase activity after this time with HCl 1N. Aliquots of the different incubations at both t=0 and t=48 h were subjected to 12% polyacrylamide gel electrophoresis under reducing and denaturant conditions to observe the potential effect of caspase proteolytic action on bovine and porcine myofibrilar proteins.

Protein identification by in-gel trypsin digestion and LC-ESI-MS/MS

Those protein bands that were effectively hydrolysed by the action of either caspase 3 or 7, together with the polypeptide fragments that were generated after 48 h incubation with caspases, were excised from the gel and digested with trypsin. After this, generated peptides were desalted using zip-tips™ and finally identified by using a Surveyor LC system coupled to a LCQ Advantage Ion trap MS instrument (Thermo Scientific). Interpretation or raw MS/MS spectra was performed using the Mascot v2.3 search engine against NCBInr protein database.

Results and discussion

The proteolytic action of recombinant caspases 3 and 7 on muscle structural proteins coming from porcine and bovine species is shown in Figure 1. As observed from the different protein electrophoretic profiles, there are some remarkable changes between myofibrillar extracts taken at t=0 and those taken after 48 hours of incubation with either caspase 3 or caspase 7.

Between the main changes we can observe a selective proteolytic action on myosin heavy chain for the two animal species, since this protein was effectively hydrolyzed by caspase 7 but not by caspase 3 (Figure 1, protein bands 1-4). Despite this evident degradation, probably the main hallmark of caspase-mediated proteolysis on myofibrillar proteins was the generation of a

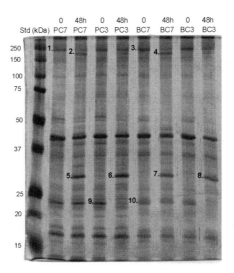

Figure 1. SDS-PAGE. 12% SDS-PAGE of the protein degradation profile obtained when porcine (P) or bovine (B) muscle myofibrillar extracts were incubated in presence of 60 nM of purified recombinant caspase 3 (C3) or caspase 7(C7) for a period of 0 (controls) or 48 hours.. Std: protein standards.

polypeptide of around 30 kDa from both bovine and porcine muscle extracts in the presence of either caspase 3 or 7 (protein bands 5-8). Proteomic analysis of these bands (Table 1) revealed that this fragment was the product of actin degradation in all cases.

According to previous research, one of the main structural changes occurring in *post mortem* muscle is the breakdown of the acto-myosin complex, contributing to the decrease the mechanical resistance of muscle and thus to the development of meat tenderness (Kemp *et al.* 2010). Our findings support the idea that, after apoptosis onset immediately after slaughter, executioner caspases 3 and 7 could contribute to muscle cell dismantling through degradation of the acto-myosin complex.

On the other hand, a protein band of around 22 kDa was observed to be highly hydrolysed by caspase 3 in both bovine and porcine muscle extracts (Figure 1, protein bands 9-10), which was not the case for caspase 7. Proteomic analysis of in-gel digests identified these bands as porcine and bovine myosin light chain 1, respectively (Table 1).

Results obtained in this work suggest a coordinated and complementary action between caspases 3 and 7 during cell dismantling. This represents an advance in exploring the consequences that apoptosis can have in postmortem muscle in relation to caspase proteolytic action, selective degradation of the muscle protein structure, development of meat tenderness and final meat quality.

Table 1. Proteins identified after trypsin digestion and LC-ESI-MS/MS analysis of the selected marked bands of Figure 1.

Band	Protein identifier (NCBInr)	Protein name	Species origin
1	gi|157279731	Myosin-1	*Sus scrofa*
2	gi|157279731	Myosin-1 (fragment)	*Sus scrofa*
3	gi|41386691	Myosin-1	*Bos taurus*
4	gi|41386691	Myosin-1 (fragment)	*Bos taurus*
5	gi|268607671	Muscle Actin (fragment)	*Sus scrofa*
6	gi|268607671	Muscle Actin (fragment)	*Sus scrofa*
7	gi|27819614	Muscle Actin (fragment)	*Bos taurus*
8	gi|27819614	Muscle Actin (fragment)	*Bos taurus*
9	gi|117660874	Myosin Light Chain 1	*Sus scrofa*
10	gi|118601750	Myosin Light Chain 1/3	*Bos taurus*

Acknowledgements

This work was financially supported by grant AGL2009-12992 from Spanish Ministry of Economy and Competitiveness.

References

D'Alessandro, A., Rinalducci, S., Marrocco, C., Zolla, V., Napolitano, F., & Zolla, L. 2012, 'Love me tender: An Omics window on the bovine meat tenderness network', Journal of Proteomics, 75, 4360-4380.

Kemp, C. M., Sensky, P. L., Bardsley, R. G., Buttery, P. J., & Parr, T. 2010, Tenderness – an enzymatic view, Meat Science, 84, 248-256.

Ouali, A., Herrera-Mendez, C. H., Coulis, G., Becila, S., Boudjellal, A., Aubry, L., & Sentandreu, M. A. 2006, Revisiting the conversion of muscle into meat and the underlying mechanisms, Meat Science, 74, 44-58.

Allergen characterization in Mediterranean farmed fish: a proteomic approach

I.M. Luís[1], A. Soggiu[2], D. Schrama[1], J. Dias[1], S. Prates[3], P. Roncada[4], L. Bonizzi[2] and P.M. Rodrigues[1]

[1]CCMAR, Universidade do Algarve, FCT, Edifício 7, Campus de Gambelas, 8005-139, Faro, Portugal; pmrodrig@ualg.pt
[2]Dipartimento di scienze veterinarie e sanità pubblica, Università degli studi di Milano, Milano, Italy
[3]Serviço de Imunoalergologia, Hospital Dona Estefânia, Lisboa, Portugal
[4]Istituto Sperimentale Italiano Lazzaro Spallanzani, Milano, Italy

Introduction

Food allergies are defined by FAO as 'abnormal responses of the immune system to certain food components' and they are usually triggered by proteins that naturally occur in food (Taylor, 2000). These allergies can be classified as mediated by immunoglobulin E and non-IgE mediated.

Non-IgE mediated allergies are not well characterized and seem to play a minor part in food allergies, with delayed symptoms and less severe allergic reactions (FAO/WHO, 2000; Taylor, 2000). However the IgE-mediated responses have a well-known mechanism, which is initiated immediately after the exposure to the offending food and can produce a quite severe allergic response.

The IgE-mediated type of allergic reaction was observed for more than 170 different foods but there are 8 foods or groups of food that account for around 90% of all the severe allergic reactions to food. Fish are one of these eight kinds of foods (FAO/WHO, 2000; Taylor, 2000).

Fish allergy is observed in 0,1 to 0,2% of the population and the better well characterized fish allergen is parvalbumin. This protein is found in high amounts in the sarcoplasmatic fraction of white muscle of fish and it is a calcium-binding protein with 12kDa and acidic isoelectric point (Carrera et al.; Chen et al., 2006; Misnan et al., 2008). Parvalbumins from different species of fish show high homology with the most conservative regions being those that form the two calcium biding sites (Chen et al., 2006) and this might be the reason why patients allergic to fish tend to develop allergic reactions with numerous fish species (Carrera et al.).

Despite being the major antigens found in fish, parvalbumins are not the only allergen present in fish. In fact, there are some other documented proteins that are able to trigger an allergic response. Aldehyde phosphatase dehydrogenase (Das Dores et al., 2002) was found to be an allergen present in cod, collagen (Hamada et al., 2001) and a fish allergen present in fish muscle and skin, the hormone vitellogenin (Perez-Gordo *et al.*, 2008) present in fish raw caviar was characterized as an antigen present in fish and some other fish proteins were found to have antigen properties although they have not been well-characterized yet (Misnan *et al.*, 2008).

Methodology and results

An experimental approach was designed in order to better characterize some Mediterranean fishes, either from wild or farming origin regarding their allergenic potential, as well as their allergen identification. This design relies on the antigen-specific IgE-mediated allergic reaction.

The identification of fish allergens was obtained by the immunodetection of proteins through one and two dimensional western blot (WB) and mass spectrometry (MS) techniques. It is possible to establish a parallel between the WB procedure and the allergic reaction after sensitization, where the common elements are the IgE antibodies present in the blood and the specific cross-linkage between antibodies and antigens, which take part in the experimental procedure through the use of sera from allergic patients.

In order to test the allergenic potential from Major Mediterranean fishes (wild and farmed) and to collect sera for the experiment, skin-prick tests were conducted in patients with fish allergy history followed in Immunoallergology Unit – *Hospital Dona Estefânia* (Lisbon, Portugal). Sera was also collected both from patients with a clinical history and non-allergic individuals as control.

Fish used for the skin-prick tests were wild and farmed Gilthead sea bream (*Sparus aurata*), European sea bass (*Dicentrarchus labrax*), Senegalese sole (*Solea senegalensis*), Meagre (*Argyrosomus regius*) and wild White bream (*Sargus sargus*). The skin-prick tests were conducted by pricking the skin on the volar surface of the forearm through a portion of fish muscle with a metallic lancet with 1mm penetration. Saline was used as negative control and histamine (10 mg/ml) as positive control. The results were read after 15 to 20 minutes. A negative reaction is characterized for no alterations in the appearance of the skin. The positive reaction is characterized by a pruritic wheal and flare reaction consisting on a white raised central swelling surrounded by a red area. The wheals are outlined with a pen and the results transferred to paper with transparent tape. The mean diameter of the wheals is measured and considered positive if it is at least 3 mm higher than the positive control. Results show that farmed fishes have higher allergenic content than the wild ones with farmed European Sea bass showing the stronger allergenic reactivity. Serum from the patients with positive skin prick test reactions was collected and kept at -80 °C.

The same fish used for the skin-prick tests were used to determine the allergenic potential through the linkage of anti-cod parvalbumin antibody and parvalbumins in tested fish. This test was performed with Fish-Chek ELISA kit (R6009-1E) acquired to Bio-Check (UK). This kit provided a plate coated with anti-cod parvalbumin that linked the parvalbumins present in fish that were then measured through the detection with the same antibody conjugated with horseradish peroxidase enzyme.

Both results obtained with the skin-prick test and ELISA indicated that European sea bass, wild and farmed, are the fish with highest allergenic reactivity and the highest content of parvalbumins. For this reason farmed European sea bass was the fish chosen to perform antigen identification.

For the immunodetection through WB, the sarcoplasmatic fraction was extracted from the white muscle of European sea bass using the method described by Kjaersgard and Jensen (Kjaersgard and Jessen, 2004) validated by Silva et al. (Silva et al., 2010). After the extraction, part of the extract was cleaned using the Ready Prep 2-D Cleanup kit (Bio-rad) and rehydrated in buffer with 8M Urea, 4% CHAPS, 1% DTT and 0,5% Ampholine pH 3,5-10.

Cleaned extract was used to test and optimize 2-DE separation. Cup loading rehydration was chosen as it provided a better 2-DE gel resolution with a higher number of visible spots (Figure 1d). Both fractions of cleaned and non-cleaned extract were used to perform 1-DE

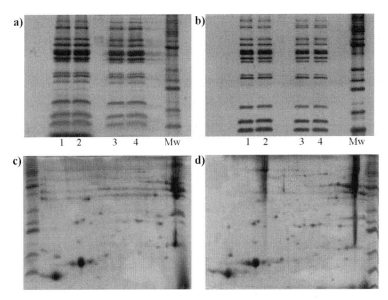

Figure 1. (A,B) 1D-SDS-PAGE gels of fish muscle extract (14%T), (C,D) Two dimensional electrophoresis (14%T) of cleaned fish muscle extract stained with colloidal Coomassie. (a) Colloidal Coomassie Blue stained gel of cleaned protein extract: lanes 1, 2; before cleanup: lanes 3,4; mw: molecular weight markers (b) silver stained gel of cleaned protein extract: lanes 1, 2; before cleanup: lanes 3,4; mw: molecular weight marker) for Coomassie Blue and Silver Staining respectively (c) passive rehydration (d) cup loading pH gradient 4 to 7 from left to right.

separation with minor differences as shown in Figure 1 (a,b) that were then used to perform preliminary tests of the WB design. The WB procedure was carried performing semi-dry transfer to Nitrocellulose membrane (0.45 μm pore) with the transfer efficiency proven by Ponceu S staining. The membrane was first incubated for 1-hour with sera from allergic or non-allergic patients (dilution 1:500), then with biotinylated anti-human IgE antibody (dilution 1:2,000) and finally with streptavidin conjugated either with alkaline phosphatase or Cy3 (dilution 1:1000).

Discussion

Despite the use of both types of anti-IgE antibodies (monoclonal and polyclonal), different sera dilutions (from 1:50 to1:500) and fluorescence revelation of blotted proteins. Unfortunately it was not possible to perform antigen detection, probably because we are under the detection limit of the method used. We were only able to detect low non-specific immune responses that according to a screening test is independent both of using allergic or non-allergic sera, as well as sera incubation We will test lower sera dilutions (1:2 to 1:10) since IgE concentrations in sera can be quite low (from 0.0001 to 0.0002 mg/ml)

WB optimization is now underway in order to detect proteins responsible for the allergic content of European Sea bass. In order to succeed we first intend to quantify sera IgE content and eventually test the direct interaction between sera and anti-human IgE.

Acknowledgements

Inês Matias Luís acknowledges support of Cost action FA 1002 – Farm Animal Proteomics, through a STSM at Prof. Paola Roncada laboratory.

References

Carrera, M., Cañas, B. and Gallardo, J.M., Proteomics for the assessment of quality and safety of fishery products. Food Research International.

Chen, L.Y., Hefle, S.L., Taylor, S.L., Swoboda, I. and Goodman, R.E., 2006. Detecting fish parvalbumin with commercial mouse monoclonal anti-frog parvalbumin IgG. Journal of Agricultural and Food Chemistry 54: 5577-5582.

Das Dores, S., Chopin, C., Romano, A., Galland-Irmouli, A.V., Quaratino, D., Pascual, C., Fleurence, J. and Gueant, J.L., 2002. IgE-binding and cross-reactivity of a new 41 kDa allergen of codfish. Allergy 57 Suppl 72: 84-87.

FAO/WHO, 2000. Assessment of the allergenicity of genetically modified foods, Food and Agriculture Organization of the United Nations, World Health Organization, Geneva, Switzerland.

Hamada, Y., Nagashima, Y. and Shiomi, K., 2001. Identification of collagen as a new fish allergen. Biosci Biotechnol Biochem 65: 285-291.

Kjaersgard, I.V.H. and Jessen, F., 2004. Two-dimensional gel electrophoresis detection of protein oxidation in fresh and tainted rainbow trout muscle. Journal of Agricultural and Food Chemistry 52: 7101-7107.

Misnan, R., Murad, S., Jones, M., Taylor, G., Rahman, D., Arip, M., Abdullah, N. and Mohamed, J., 2008. Identification of the Major Allergens of Indian Scad (*Decapterus russelli*). Asian Pacific Journal of Allergy and Immunology 26: 191-198.

Perez-Gordo, M., Sanchez-Garcia, S., Cases, B., Pastor, C., Vivanco, F. and Cuesta-Herranz, J., 2008. Identification of vitellogenin as an allergen in Beluga caviar allergy. Allergy 63: 479-480.

Silva, T.S., Cordeiro, O., Jessen, F., Dias, J. and Rodrigues, P.M., 2010. Reproducibility of a fractionation procedure for fish muscle proteomics. American Biotechnology Laboratory 28: 8-13.

Taylor, S.L., 2000. Emerging problems with food allergens. Food, Nutrition and Agriculture.

Protein exudation from salted meat in processing by cold tumbling

Priit Soosaar[1,2], Avo Karus[2], Meili Rei[2] and Virge Karus[2]
[1]*UAB EFIS Eesti Filiaal, Tartu, Estonia*
[2]*Institute of Veterinary Medicine and Animal Science, Estonian University of Life Sciences, Tartu, Estonia; avo.karus@emu.ee*

Objectives

Protein solubilisation during preliminary steps to prepare smoked ham has several effects on product quality. Firstly, it makes the meat more sticky, which increases the outcome and quality of the final product. Secondly, it has a negative effect as some of the proteins, and consequently some nutritional value is lost. Proteomics can add value to meat technology research providing fundamental basis and thereof to optimize the technology (Bendixen, 2005).

The aim of the study was to investigate the dynamics of protein extraction from meat during cold tumbling and to establish optimal duration of the process to avoid protein lost in ham production.

Material and methods

Pork and beef were used to investigate protein solubilisation. Meat pieces were salted by injection of 7% salt (density at 4 °C 7.1 °Be) in an amount of 30% of meat weight. Tumbling was performed in repeated 10 min tumbling and 20 min rest step in duration of 18 hours at 0.9 bar and 0 °C. Total tumbler processing time was 360 min for each batch. Liquid exudates were collected at six hours, 12 hours and 18 hours respectively from five different locations in tumbler.

SDS-PAGE with 3% stacking and 8% running gel was used for characterization of meat protein exudates. Electrophoresis was performed in blocks (Himifil, Estonia) at 4 °C pH 8.3 using 2 mA per sample in stacking and 5 mA per sample in running gel. Proteins were stained with Coomassie and scanned using *Zeineh Soft Laser Scanning Densitometer*. Marker proteins from Sigma (USA) were used. Seven main protein fractions were identified and analysed:
Fraction I – M>71,000 Da, including proteins like myosin, actomyosin etc.;
Fraction II – 64,000 Da < M <71,000 Da, including tropomyosin;
Fraction III – 56,500 Da < M < 64,000 Da;
Fraction IV – 51,000 Da < M < 56,500 Da;
Fraction V – 41,000 Da < M < 51,000 Da, including G-actin;
Fraction VI – 22,000 Da < M < 41,000 Da, including troponin I and troponin T;
Fraction VII – M < 22,000 Da, including myoglobin and troponin C.

Results and discussion

Since the ionic strength of injected salt solution was relatively high, it was assumed that all observed changes are mainly due to processes directly in the studied matrix and only slightly affected by osmosis. As demonstrated (Barbieri and Rivaldi, 2008), major changes in exudates protein content are observed during the first six hours of tumbling. In tumbling during first stage salt soluble proteins are partly extracted and in a later phase meat tissue cellular disruption will take place and by salt introduced swelling of myofibrils occurs. In our study we focused only on later stages to investigate main processes in liquid exudates.

The liquid of pork meat processing batches shows significant decrease of fractions I and II (high molecular mass proteins) relative content 6 to 18 hours processing. However, the relative content of fraction VII (low molecular weight proteins) will significantly increase only from 12 to 18 hours processing (Table 1). This can be partly due to the destruction of bigger proteins into smaller fractions. In beef batches (Figure 1) significant decrease of fraction III from 6 to 12 hours was measured.

However, from 12 to 18 hours this fraction tends to increase again and fraction IV decreases. In summary, for all batches, the highest proportion in liquid tumbling exudates were protein fractions with molecular weights ranging 22-41 kDa.

It is important to mention that in beef meat batches the small protein fraction (VII) relative content at the end of tumbling was higher than in pork processing. Proteins with high molecular mass tend to decrease during tumbling and concentration of small proteins and peptides tends to increase. This effect may be only partly explainable by protein degradation, because the increase of small protein fractions is not enough to account for the decrease of big proteins fractions. Another reason of decrease of fractions I and II can be the fact that

Table 1. Relative protein concentration in fractions (average ±SD; n=10).[1]

Fraction	Tumbling time 6 h	Tumbling time 12 h	Tumbling time 18 h
I	0.11±0.04[a]	0.10±0.03[b]	0.09±0.02[b]
II	0.17±0.05[a]	0.16±0.04[b]	0.16±0.04[b]
III	0.18±0.05	0.18±0.05	0.18±0.06
IV	0.10±0.04	0.10±0.03	0.10±0.04
V	0.05±0.03	0.06±0.03	0.06±0.04
VI	0.23±0.09	0.23±0.07	0.22±0.09
VII	0.17±0.03[a]	0.18±0.07[a]	0.19±0.07[b]

[1] Values with different superscripts are significantly different; $P<0.05$.

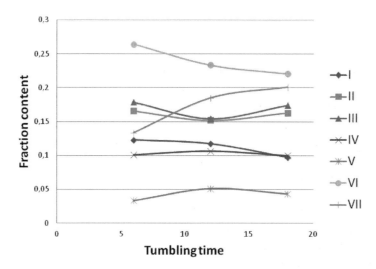

Figure 1. Protein fractions in liquid extract by beef tumbling. Fractions: I – M>71,000 Da; II – 64,000 Da < M <71,000 Da; III – 56,500 Da < M < 64,000 Da; IV – 51,000 Da < M < 56,500 Da; V – 41,000 Da < M < 51,000 Da; VI – 22,000 Da < M < 41,000 Da; VII – M < 22,000 Da (averages; n=10).

they may be more easily denatured and participate in foam-forming and therefore have lost or reduced their solubility.

The 2D analysis of proteome and identification of precise changes has to be continued.

Acknowledgements

Work was performed in collaboration with Tartu MF and AS Rey.

References

Barbieri, G. and Rivaldi, P., 2008. The behaviour of the protein complex throughout the technological process in the production of cooked cold meats. Meat Sci 80: 1132-1137.

Bendixen, E., 2005. The use of proteomics in meat science. Meat Sci 71: 138-149.

Peptidomics in extended shelf life dairy products, for assessment of proteolysis and product quality

Nanna Stengaard Villumsen[1], Marianne Hammershøj[1] Lotte Bach Larsen[1], Line Ravn Nielsen[2], Kristian Raaby Poulsen[2] and John Sørensen[2]
[1]Department of Food science, Aarhus University, Blichers allé 20, 8830 Tjele, Denmark; Nannas.villumsen@agrsci.dk
[2]Arla Foods Ingredients, Sønderrupvej26, 6920 Videbæk, Denmark

Objectives

Unwanted protein aggregation

Some extended shelf life dairy products develop insoluble material which is referred to as unwanted protein aggregation. This phenomenon appears at varying time points during the shelf life. It occurs seemingly randomly in some productions and not in others for the same product. So far no decisive factor for its development has been identified, indicating protein chemical based differences between the products, due to biological diversities in the applied raw material. This is a very plausible assumption since milk is supplied from multiple farms to the dairies. In this particular project we used re-hydrated industrial whey protein isolate (WPI) produced from cheese whey. This originates from multiple cheese dairies, which also adds to the inhomogeneity of the product.

Proteolysis and the 'protease web'

Cow milk contains multiple proteases, many of which are vital to the calf for digestion of the milk proteins, and in addition some of these play important roles during cheese production and ripening. However in extended shelf life dairy products heat stable proteases – having survived the preceding heat treatment process – can cause quality problems in the products. Proteolytic cleavage of a protein can change its solubility, surface structure, and charge; which in time might initiate the unwanted protein aggregation.

The 'protease web' in dairy products is very complex. The enzymes can be bovine like Plasmin or Cathepsin B, D or G, or originate from microorganisms in which case many are very unspecific and poorly characterized. Furthermore the proteases can be industrially introduced to the products like Chymosin and proteases from starter cultures in cheese production. Many proteases exhibit partially overlapping though distinct substrate profiles, meaning that many substrates can be cleaved by multiple proteases. To add to the complexity, proteases interact directly or indirectly with other proteases as well as their activators and inhibitors. Finally the relative amounts of the individual proteases, their substrates, activators and inhibitors, have a great impact on the proteolytic activity (Schilling and Overall, 2007).

In the dairy industry the products from silo milk or whey cover large biological diversities. Differences in the content, level and activity of the heat stable proteases, their substrates, activators and inhibitors might be present. Genetic variances in the cows could change cleavage sites and protease exosites involved in substrate recognition and cleavage, and hence result in varying proteolysis of the milk proteins. Hence there are many parameters which could potentially be connected to the observed quality problems and the aim of this project is to locate and describe this connection.

Proteolytic assays vs. peptidomics

Due to the heterogeneity of the raw material, conventional proteolytic assays, including biased searches for specific protease activities, might not be the most applicable approach. The large biological diversity of the raw material and the synergies between various proteases would be hard to simulate. Furthermore, as the aggregation can take months to form, the potential proteolysis activity is likely to be rather low – and therefore beneath the detection limit of conventional proteolytic assays.

Therefore a peptidomic approach has been chosen instead. The strategy is to map the peptide fraction present in the products with varying degrees of unwanted protein aggregation. All things being equal, differences in the protease systems will lead to differences in the observed peptide profile of a given product. The amino acid sequence of the identified peptides combined with knowledge about milk protein cleavage sites can generate hypotheses about the responsible proteases role in the generation of the individual peptides which hopefully in time can bring forth new knowledge about protease interactions in dairy products. Furthermore unique peptide profiles could be used as reference maps for the different quality parameters of specific products. This would be an easy and reliable method for prediction of quality and shelf life at an early point in production, and would therefore provide the dairy industry with a valuable tool.

Material and methods

Heat treatment

Four productions of the same whey protein isolate known to display two different degrees of unwanted protein aggregation were dissolved at 7% and adjusted to pH 3. The products were then subjected to a heat treatment of 95 °C for 180 sec, in a laboratory water bath, and then stored at -18 °C until the day of the experiment.

Peptide fractionation with centrifugal filter units and handmade RP columns

All samples were thawed simultaneously and the peptide fractions was isolated using Centrifugal filter units from Amicon®Ultra with a cutoff of 3kDa and 10kDa respectively. The samples were spun at 13000g for 15min and the permeates were desalted and concentrated on handmade chromatographic columns and then eluted onto a MALDI target plate in a volume of ~1μl, with α-cyano-4hydroxycinnamic acid in 50% acetonitril as described by (Larsen *et al.*, 2010)

Peptide identification with MALDI TOF MS and MS/MS

A Bruker Ultraflex MALDI-TOF tandem mass spectrometer (Bruker Daltonik GmbH, Bremen, Germany), was used in reflector mode to perform mass analysis of the peptide fraction. The monoisotopic masses of 7 standard peptides ranging in molecular mass from 1,046.54 to 3,147.47, was used to calibrate the instrument.

Dominant peaks and peaks with m/z corresponding to already reported milk protein peptides were further fragmented with MS/MS followed by mass searches against the SwissProt Database (Swiss Institute of Bioinformatics, Geneva, Switzerland), using the ion search program Mascot (Matrix Science, Boston, MA). The search parameters were 'none' for enzyme, '1' allowed missed cleavage, error tolerance of 100ppM for parent ion mass and 0.5 Da for fragment ion masses.

Results and discussion

The spectra identified multiple peptides from different milk proteins. A subset of these is listed in Table 1. All identified peptides were found to originate from the caseins, which are very loosely structured and highly susceptible to proteolytic degradation. Since all the casein proteins – except from cGMP – are removed from the whey fraction in the cheese production, the casein derived peptides must have been generated prior to or during this process. The κ-casein derived peptides could also be due to secondary hydrolysis of cGMP after the initial cleavage by Chymosin of the 105-106 peptide bond during cheese production. Chymosin has been shown to produce several different fragments of cGMP at low pH, (Julian R. Reid, 1997) and many of these were identified in the spectra. As the chymosin proteases used at Danish cheese dairies are heat labile, these fragments were most likely produced prior to the heat treatment; however a low chymosin activity in the products after the heat treatment cannot be ruled out without further studies. Storage experiments with the re-dissolved WPIs combined with quantitative MS would reveal whether any of these fragments increase in concentration over time, and potentially some of these data could explain and/or predict the formation of unwanted protein aggregation in the long shelf life products.

The fact that no peptides from whey proteins were detected in the rehydrated WPIs, was surprising. Even though whey proteins are generally quite resistant to proteolytic degradation due to their globular conformation, it was suspected that the heat treatment which the products have undergone would have made them more susceptible to proteolysis. More sensitive analyses including LC pre fractionation of the peptides might reveal whey protein derived pep ions for the proteases having generated specific peptides. The N-terminal Leu[191] in β-casein

Table 1. Identified peptides, and proteases suggested of being responsible for their generation.

Peptide	M/Z	Sequence	N-terminal protease	C-terminal protease	References
α-s1-casein					
1-23	2,763	RPKHPIKHQGLPQEVLNENLLRF	No cleavage[1]	Cathepsin B/D	Hurley et al., 2000, Considine et al., 2004
2-23	2,608	PKHPIKHQGLPQEVLNENLLRF	AP[2]	Cathepsin D	Hurley et al., 2000, Considine et al., 2004
91-100	1,266.7	YLGYLEQLLR	Plasmin	Plasmin	Le Bars D, Gripon JC, 1993
α-s2-casein					
198-207	1,251	TKVIPYVRYL	Plasmin	No cleavage	Le Bars D, Gripon JC, 1989
β-casein					
163-190	3,137.6	LSLSQSKVLPVPQKAVPYPQRDMPIQAF	Cathepsin D + AP	Elastase	Hurley et al., 2000
163-191	3,250.7	LSLSQSKVLPVPQKAVPYPQRDMPIQAFL	Cathepsin D + AP	Cathepsin D	Hurley et al., 2000
191-207	1,895	LLYQEPVLGPVRGPFPI	Cathepsin G/Elastase	CP[3]	Hurley et al., 2000
193-206	1,555.8	YQEPVLGPVRGPFP	Cathepsin D	CP	Hurley et al., 2000
193-207	1,668.9	YQEPVLGPVRGPFPI	Cathepsin D	CP	Hurley et al., 2000
193-208	1,781.9	YQEPVLGPVRGPFPII	Cathepsin D	CP	Hurley et al., 2000
193-209	1,880.9	YQEPVLGPVRGPFPIIV	Cathepsin D, Lb. Helv. NCC2765	No cleavage	Hurley et al.2000, Robert et al., 2004
58-76	2,060.4	LVYPFPGPIPNSLPQNIPP	Thermolysin	Thermolysin	Otte et al., 2007
191-209	2,107	LLYQQPVLGPVRGPFPIIV	Lb. Helv. CP790	No cleavage	Yamamoto and Tankano, 1999
κ-casein					
106-124	2,139	MAIPPKKNQDKTEIPTINT	Chymosin	Chymosin	Reid et al, 1997
161-169	905	TVQVTSTAV	Chymosin	Chymosin	Reid et al, 1997
162-169	803	VQVTSTAV	Chymosin	Chymosin	Reid et al, 1997

[1] No cleavage indicates that the peptide is derived from the very N or C- terminal and therefore not the result of proteolytic cleavage.

[2] AP = Aminopeptidase.

[3] CP = Carboxypeptidase.

derived peptides, may originate from the bovine proteases Cathepsin G or Elastase (M. J Hurley, 2000) but also from microbial proteolysis from *Lactobacillus helveticus*,(Robert *et al.*, 2004). As the protease web includes different proteases with shared substrate specificities this is definitely plausible, and further studies including proteolytic assays are needed to determine the proteases responsible for the observed peptides.

References

Considine T., H.Á., Kelly A.L., McSweeney P.L.H., 2004. Hydrolysis of bovine caseins by cathepsin B, a cysteine proteinase indigenous to milk. International Dairy Journal 14: 117-124.

Hurley, M.J, Larsen, L.B., Kelly, A.L and McSweeney, P.L.H., 2000. Cathepsin D activity in quarg. International Dairy Journal 10: 453-458.

Julian R. Reid, T.C., John S. Ayers, Kate P. Coolbear, 1997. The action of chymosin on κ-casein and its macropeptide: Effect of pH and analysis of products of secondary hydrolysis. International Dairy Journal 7: 559-569.

Larsen, L.B., Hinz, K., Jorgensen, A.L., Moller, H.S., Wellnitz, O., Bruckmaier, R.M. and Kelly, A.L., 2010. Proteomic and peptidomic study of proteolysis in quarter milk after infusion with lipoteichoic acid from *Staphylococcus aureus*. J Dairy Sci 93: 5613-5626.

Le Bars, D. and Gripon, J.C., 1989. Specificity of plasmin towards bovine alpha S2-casein. J Dairy Res 56: 817-821.

Le Bars, D. and Gripon, J.C., 1993. Hydrolysis of αs1-casein by bovine plasmin. Lait 73: 337-344.

Otte J, S.S., Zakora M, Nielsen MS, 2007. Fractionation and identification of ACE-inhibitory peptides from α-lactalbumin and β-casein produced by thermolysin-catalysed hydrolysis. International Dairy Journal 17: 1460-1147.

Robert, M.C., Razaname, A., Mutter, M. and Juillerat, M.A., 2004. Identification of angiotensin-I-converting enzyme inhibitory peptides derived from sodium caseinate hydrolysates produced by Lactobacillus helveticus NCC 2765. J Agric Food Chem 52: 6923-6931.

Schilling, O. and Overall, C.M., 2007. Proteomic discovery of protease substrates. Curr Opin Chem Biol 11: 36-45.

Yamamoto, N. and Takano, T., 1999. Antihypertensive peptides derived from milk proteins. Nahrung 43: 159-164.

Egg proteome: the protein and carotenoids characteristics deriving from laying hens reared in different housing systems and under different environmental conditions

Charles Spiteri[1], Everaldo Attard[1], Cesare Castellini[2] and Adrian Bugeja Douglas[1]
[1]*University of Malta, Institute of Earth Systems, Division of Rural Sciences and Food Systems, Malta; everaldo.attard@um.edu.mt*
[2]*Universita' Degli Studi di Perugia, Dip Applied Biology, Perugia, Italy*

Objectives

The rearing and breeding of poultry both for meat and egg production has long been an important part of the Maltese agricultural activity. In fact, it can safely be said that historical evidence affirm the highly importance of poultry in the hands of Maltese farmers. This can nowadays be proofed by the presence of a local rustic dual purpose poultry breed known as the 'Black Maltese'. Nowadays local egg production has been entirely replaced by intensively reared hybrid strains of poultry, particularly those deriving from the White Leghorn and Rhode Island Red, both of which are more productive.

We attempted to determine the egg characteristics in conformance with COUNCIL DIRECTIVE 1999/74/EC.

Material and methods

This research was based on the assessment and examination of the physical and chemical characteristics of eggs derived from two different housing systems for laying hens whilst taking also into consideration thermal conditions and flora availability for the free-range (FR) hens. The housing systems used for this research were the free-range and the enriched cage (EC) systems. Physical characteristics included weight and measurement of eggs, and chemical parameters; carotenoids in yolk and protein in albumen. Eight egg samples were analyzed each time; 4 from FR and the other 4 from EC population. This was carried out for 16 times, with a total of 128 eggs were tested.

The carotenoid content in the eggs was evaluated by a new system that was recently developed by BioAnalyt GmbH (Teltow, Germany). Briefly, 500 mg of yolk was taken from each egg for both the FR and EC egg types. These were pooled in samples of two. 400 mg of egg yolk was taken from the pooled sample and diluted to a final weight of 2.00 g with dilution buffer. The diluted sample was adequately shaken so that the egg yolk sample mixes well with the buffer solution. 400 µl of the diluted egg yolk was injected into the extraction (iEx, BioAnalyt GmbH, Germany) vial. Following 10 sec shaking, the vials were left for complete phase separation, approximately 5 min. The concentration of carotenoids in the upper organic phase was then measured using the portable photometer (iCheck™, BioAnalyt GmbH, Germany) and final concentration (mg/kg) was calculated based on the exact sample weight and final buffer weight.

The protein content was evaluated using a colorimetric-densitometric analysis. Briefly, 1000 µl of albumen from each egg was diluted in 9ml of distilled water. 10 µl from the diluted samples was extracted and applied to the silica gel plates (Whatman flexible Plates, Aldrich, Germany) in triplicates. The extracted samples were allowed to dry in air and then sprayed with the ninhydrin reagent (60mg ninhydrin, 20ml n-butanol and 0.6ml acetic acid; Wagner, and Bladt, 2001). The silica gel plates were then dried at 105 °C in an oven for 5 minutes. After drying, the plates were scanned and processed densitometrically by ImageJ software. The scanned images were split into three colours (red, blue and green) and the red channel was used for the analysis. The grey scales of the image were inverted so as to analyse the spots against a dark background. A 52×52 pixel square was selected and background/sample images were measured. The mean background grey value was approximately 0. The background noise generated by the staining of the background with the ninhydrin spraying reagent was eliminated.

Results and discussion

From egg weight analysis it was observed that until the 16th day of sampling, there was a statistical significant difference ($P<0.05$, v=16) between the values obtained. The eggs deriving from the free-range hens were heavier (67.41 g) than those deriving from hens in enriched cages (63.61 g). Parameters which may have caused this difference in values can be nutrition and production. However it was noted that hens reared in the enriched cages deposited eggs with reduced weight at a higher production laying rate. This was also mentioned by Leeson and Summers (2000). From egg measurements observations, although the free-range egg measurements were at times bigger than those of enriched caged hens there was no significant difference between the two types. The free-range laying hens did not deposit eggs daily as the enriched caged hens did.

Values for the carotenoid content in the egg yolks clearly show that the free-range eggs had a significantly higher carotenoid content (20.18 mg/kg) than those eggs deriving from the enriched caged hens (13.13 mg/kg). This is reflected through the feed intake. The hens were fed the same mashed diet, however the free-range hens also foraged on available flora during this period and possibly even insects. Plants included cape sorrell (*Oxalis pes-caprae*), clustered sulla (*Hedysarium glomeratum*), common Birdsfoot Trefoil (*Lotus edulis*), crown daisy (*Chrysanthemum coronarium*), smooth sow thistle (*Sonchus Oleraceus*) and the white mustard (*Sinapis alba*). Results confirm that carotenoids added more colour to the yolk of the free-range eggs. Moreover, the cape sorrel was the preferred yellow-petaled flower. Traditionally it was used as part of nutrition during late winter and spring to stimulate egg laying when concentrated livestock feed was not easily available.

The protein content in both egg samples were not statistically different although notable differences when observed. In all cases it was observed that the mean protein index in FR eggs (65.34) was slightly higher than that for the EC eggs (58.23). On some days, the protein index of EC eggs was higher than that of the FR eggs, coinciding with an abrupt rise in temperature during late spring and early summer (Figure 1). This is possibly explained by the fact that the FR hens, were more exposed to the outside hot climatic conditions and hence consumed more water than the EC hens. Consequently, the water content of the egg may have increased.

Whereas differences in carotenoid content, between the two groups, were greatly influenced by the climatic conditions and the presence of particular vegetation, no differences in protein contents were observed by such parameters. However, both groups showed a similar trend in protein contents with variations in the daily temperature-humidity index.

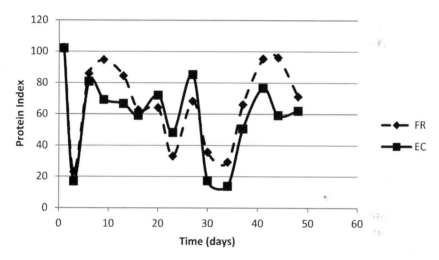

Figure 1. Protein index against time.

Acknowledgements

We would like to thank BioAnalyt GmbH who provided us with the test kit and instrumentation to analyse the carotenoid content of the eggs.

References

COUNCIL DIRECTIVE 1999/74/EC of 19 July 1999 'Laying down minimum standards for the protection of laying hens.
Leeson, S. and J.D. Summers, (2000). Broiler breeder production. University books, Guelph, Ontario.
Wagner, H. and Bladt, S. 2001, Plant Drug Analysis, 2nd edition, Springer, Berlin.

Proteomics and transcriptomics investigation on longissimus muscles in Large White and Casertana pig breeds

Anna Maria Timperio , Federica Gevi, Valentina Longo, Valeria Pallotta and Lello Zolla
Laboratory of Proteomics Department of Ecology and Biology University 'La Tuscia' Viterbo Italy; zolla@unitus.it

Objectives

Consumer complaints against the blandness of modern lean meat and the frequent reference to the more strongly flavored meat that was available years ago have prompted reconsideration of high fat-depositing typical pig breeds. Casertana and Large White pig breeds are characterized by a different tendency toward fat accumulation as they exhibit opposite genetic and physiological traits with respect to the energy metabolism. These physiological differences were investigated in longissimus lumborum muscles through proteomics (2-DE, MS/MS) and microarray approaches. Data were analyzed for pathway and network analyses, as well as GO term enrichment of biological functions. As a result, Casertana showed a greater amount of proteins involved in glycolitic metabolism and mainly rely on fast-mobilizable energy sources. Large White overexpressed cell cycle and skeletal muscle growth related genes. Metabolic behavior and other implications are discussed.

Material and methods

Sample collection

All animals used in this study were treated according to International Guiding Principles for Biomedical Research Involving Animals. In a commercialdairy farm located in Viterbo (Italy), we selected 20 animals (10 per breed) which were of the same age and fed the same diet.

Sample preparation and proteomics analysis

Sample preparation and solubilization was performed by slight modificationof the SWISS-2D PAGE sample preparation procedure (Hoogland *et al.*, 2004). Semiquantitative IEF-SDS PAGE. Eight-hundred micrograms of protein were resuspended in 7 M urea, 2 M thiourea, 4% CHAPS, pH 3-10 carrier ampholyte (total volume 250 µl) and then used to rehydrate 18 cm long IPG 3-10 NL (Amersham Biosciences) for 8 h (1 mg of protein each strip). IEF was carried out on a Multiphor II (Amersham Biosciences) with a maximum current setting of 50 µA/strip at 20 °C

Image analysis

Sixty stained gels (3 technical replicates × 10 biological replicates × 2 breeds) were digitalized using an ImageScanner and LabScan software 3.01 (Bio-Rad Hercules, CA). The 2-DE image analysis was carried out and spots were detected and quantified using the Progenesis SameSpots software v.2.0.2733.19819 software package (Nonlinear Dynamics, New Castle, UK). Protein Identification by Nano-RP-HPLC-ESI-MS/MS. Mass spectrometric procedures were performed as previously described. Peptide mixtures were separated using nanoflowHPLC system (Ultimate; Switchos; Famos; LC Packings, Amsterdam, the Netherlands)(D'Amici *et al.*, 2008).

RNA sample extraction

Longissimus lumborum muscle samples of LW and CA individuals (see above) were collected immediately after slaughtering. Samples were preserved in RNA later (Sigma-Aldrich) and stored at -80 °C.

Microarray experiments

RNA samples from the 10 animals of each breed were pooled as to reduce the total amount of needed material and because the primary interest was to put evidence on breed-specific gene expression changes rather than individuals.

Microarray data analysis

Hybridization images were scanned with the Packard ScanArray Express Line of Microarray Scanners (PerkinElmer Life Sciences, Waltham, Massachussets). Spot finder software (TIGR) was used to extract feature data from microarray fluorescence images. Slides were first pre-processed filtering spots with poor hybridization signals, saturated signals, or signal to background ratio lower than two. Then lowess and dye-swap normalizations were applied. Functional Enrichment of GO Terms. In a second approach, some of the representing proteins should be examined in databases to get additional annotations on their functions and consequently to establish some hypothesis concerning their functions in the suckling upon lactation.

Results and discussion

In the present study, we investigated protein composition of longissimus lumborum muscle samples from Large White (LW) and Casertana (CA) pig breeds. In parallel, samples were also analyzed with transcriptomics techniques. Proteomics analyses (2-DE and MS/MS identification) yielded a total of 473 (44 spots which were commonly expressed by both breeds, and 36 differentially expressed spots ($P = 0.05$) upon comparison of 60 gels (30 per breed, 10 biological replicates and 3 technical replicates each). In particular, 13 spots were up-regulated in CA and 20 in LW. Approximately 95% of these spots were identified via ESI-MS/MS. Notably, mass spectrometric identification of those spots revealed that most of those spots accounted for the same protein (mainly glycolytic enzymes in CA and myosin light chain isoforms in LW, as to yield a total of 9 different up-regulated proteins in CA and 10 counterpart entries for LW. Microarray analyses have individuated a total of 105 differentially expressed gene transcripts with a fold change >1.3. In detail, 66 gene transcripts were up-regulated in CA and 39 in LW pigs.

However, in addition to various biological factors, it should be considered that the poor correlation between transcriptomic and proteomic data could be quite possibly due to the inadequacy of available statistical tools to compensate for biases in the data collection methodologies as well as to the different bioinformatic approaches employed to analyze and integrate data from either proteomics or genomics studies. There are also difficulties in evaluating on a global level which biological factor, translational efficiency or protein half-life, influences the correlation between mRNA and protein abundances to time-course differences between mRNA changes and protein responses (Maier *et al.*, 2009). Under this perspective, proteins and gene transcripts indirectly converged in CA, as they mainly belonged to metabolic pathways – glycolytic and glycolysis-related enzymes, such as Enolase 3 (ENO3), Triosephosphate Isomerase (TPI), Phosphoglucomutase1 (PGM1), Lactate Dehydrogenase (LDHA), Glucose 3-Phosphate Dehydrogenase (GPDH) and Creatine Kinase (CK-M), Ketoexokinase (KHK) enzymes involved in (1) fatty-acid oxidation responsessglutaredoxin 3 (GRX3), thioredoxin 2 TRX2; (2) calcium homeostasis – S100 calcium binding protein A2 (S100A2); and (3) hormone inducers/growth factors/regulators of transcription (hydroxy-delta-5-steroid dehydrogenase (HSD3B2), NFKB inhibitor interacting Ras-like 1 (NKIRAS1), bromodomain containing 4 (BRD4), leucine-rich repeat kinase 1 (Lrrk1), silent mating type information regulation 2 homologue 2 (SIRT1).

Conversely, proteins and gene transcripts which were found to be overexpressed in LW longissimus muscle mainly accounted for (1) structural muscle proteinssmyosin light chain (MLC) isoforms 1f, 2, 2 V and 3; alpha-actin (ACTA1); (2) proteins involved in the maintenance of the balance between protein synthesis and degradation – E3 ubiquitin-protein ligase (MARCH5), ring finger protein 128 (RNF128), SERPINA3; (3) fatty-acid oxidation and oxidative stress responses carnitine palmitoyltransferase 1C (CPT1), Glutathione peroxidase 5 (GPRX5),Peroxiredoxin-2(PRX2); (4) transcriptionegulatorssretinoic acid receptor alpha (RARA), estrogen related receptor alpha (ESRRA), basic leucine zipper and W2 domain-containing protein 2 (BZW2), zinc finger protein 212 (ZNF212). These preliminary observations were confirmed upon GO term enrichment of biological function in CA Longissimus upregulated proteins and transcripts Indeed, CA muscles appeared to be particularly enriched in GO terms involving glycolytic catabolism.

In conclusion this study was aimed at detecting representative molecules which could be related to the increased fat deposition attitude of the Italian breed and the leanness of the meat of LW pigs. Most of the differentially expressed individuated proteins and gene transcripts account for proteins which have been proposed to play a role in meat quality in literature.

Acknowledgements

This study has been supported by the 'GENZOOT' research programme, funded by the Italian Ministry of Agriculture. The Authors would like to thank Dr. Bianca Moioli for her stimuli and support.

References

D'Amici, G.M., Timperio, A.M. and Zolla, L., 2008. Coupling of native liquid phase isoelectrofocusing and blue native polyacrylamide gel electrophoresis: a potent tool for native membrane multiprotein complex separation. Journal of Proteome Research 7: 1326-1340.

Hoogland, C., Mostaguir, K., Sanchez, J.C., Hochstrasser, D.F. and Appel, R.D., 2004. SWISS-2DPAGE, ten years later. Proteomics 4: 2352-2356.

Maier, T., Guell, M. and Serrano, L., 2009. Correlation of mRNA and protein in complex biological samples. FEBS Lett 583: 3966-3973.

Acknowledgements

The Organizing Committee of the Farm Animal Proteomics 2013 meeting acknowledges the support of the entities and individuals that made possible this publication and event.

Major sponsors

Bruker s.r.o. (www.bruker.com)

MERCK spol. s.r.o.
(www.merck-chemicals.sk)

We make it visible.
Carl Zeiss spol. s.r.o. (www.zeiss.cz)

KRD molecular technologies s.r.o.
(www.krd.cz)

Jena Bioscience GmbH
(www.jenabioscience.com)

SARSTEDT (www.sarstedt.com)

BIOTECH s.r.o. (www.biotech.cz/sk)

SAIA, n.o. (www.saia.sk)

VWR (sk.vwr.com)

ZLATÝ DUKÁT (www.zlatydukat.sk)

Organizing support

Univerzita veterinárskeho lekárstva a farmácie v Košiciach (www.uvm.sk)

COST-FAProteomics.org (www.cost-faproteomics.org)

COST (European Cooperation in Science and Technology) (www.cost.eu)

Printed in the United States
by Baker & Taylor Publisher Services